William Ward Van Valzah

The chronic disorders of the digestive tube

William Ward Van Valzah

The chronic disorders of the digestive tube

ISBN/EAN: 9783337729202

Printed in Europe, USA, Canada, Australia, Japan

Cover: Foto ©berggeist007 / pixelio.de

More available books at **www.hansebooks.com**

THE CHRONIC DISORDERS

OF THE

DIGESTIVE TUBE

BY

W. W. VAN VALZAH, A.M., M.D.
**Formerly Demonstrator of Clinical Medicine,
Jefferson Medical College**

NEW YORK
J. H. VAIL & CO.
1893

PREFACE.

Tins little book, with the exception of the chapter on habitual constipation, is made up of communications during the past year to the *Journal of the American Medical Association*, the *New York Medical Journal*, and the *Medical Record*. I have been persuaded to combine and reprint these articles under one cover, in order to present to the profession, in an easily accessible form, a short and practical study of the chronic disorders of the alimentary tract. Originally intended for serial publication, no very great changes have been found necessary to adapt them to the present form.

Great pains have been taken to make each chapter complete in itself. This plan has both its advantages and disadvantages. It relieves the busy reader of the necessity of going through the book in order to find the author's treatment of a particular disorder; but it also renders it impossible to avoid repetition of certain basic and controlling principles. The importance (in the opinion of the writer) of these principles is a satisfactory explanation and apology for their frequent statement.

Popular opinion places seasickness among the disorders of the stomach. This contention is shown to be erroneous, and an attempt is made to explain the nature of this neglected disease. A justification for its consideration under this title may be found in the fact that to secure healthy digestion and motility before and during the voyage is the best way to prevent the gastro-intestinal disturbances secondary to this peculiar sensory form of vertigo.

No. 10 EAST FORTY-THIRD STREET,

NEW YORK, December 1st, 1892.

CONTENTS.

THE

CHRONIC DISORDERS

OF THE

DIGESTIVE TUBE.

CHAPTER I.

GENERAL THERAPEUTIC CONSIDERATIONS.

THE results of the surgical treatment of disease
are palpable and often brilliant. The wonderful
achievements and rapid advances of modern surgery are manifest, and its results can be built up
into statistics that will not yield to scepticism's
destroying touch. It is not so in medicine. Our
great triumphs are in the prevention and control
as well as in the cure of disease, and the entire
good that we do cannot be known. The surgeon
believes in the knife because he sees its power,
recognizes its limitations, brings other powerful
means to its aid, and proceeds, in a way often
clearly marked out in every detail, to the accomplishment of a definite purpose. The physician's

1

scepticism is born of the obscurity of therapeutic results, faulty and narrow methods, and a failure to recognize the limitations imposed by the nature and stage of the morbid process. But we do more than we know, or are able to explain. We are powerless at the bedside only when therapeutic nihilism has boldly swept away every landmark and light. In the hour of transcendent need the physician, standing in the dim twilight, bends forward into the darkness to cure, to strengthen, and to bless.

In the chronic disorders of the digestive tube scepticism finds an almost impregnable stronghold, well barricaded by professional and popular opinion against successful assault. These troubles rarely disappear in the course of nature, for they possess an inherent power of self-perpetuation, and many physicians seem to have a little too much faith in the unaided power of drugs. These invalids too frequently fail to get the slightest benefit from the few magical prescriptions of even the best and greatest men of our profession, and turn to folk-lore for relief, and a little later a few more victims are added to the list of the consumers of the many patent medicines "good for digestion and the liver, and diarrhœa and constipation." And thus our profession falls into disrepute. It might be well to prescribe a little less medicine, and be a little more explicit and full and emphatic in orders concerning hygiene and dietetics. Drugs, by affording an excuse for the neglect of other more powerful remedies, too often become the world's grave-diggers; and this little book shall not accomplish its purpose

if it fails to establish the essential utility of right
living and a proper diet in the treatment of the
chronic disorders of digestion and nutrition. Medical practice knows no more brilliant results than
those obtained by the right management of these
diseases. But the treatment must be well defined,
comprehensive, and thoroughly and systematically
carried out. It may be well to go into detail and
define the basis, purpose, and some of the necessary
limitations of the therapeutics of these chronic disorders of the alimentary canal.

Disease is in its primitive nature a perversion
of force which determines the fixed pathological
changes and often dominates the symptomatic expression. This fundamental truth must be recognized before therapeutics can claim to be a science
rather than an art based on the contradictory testimony of experience. Curative therapeutics must
go beyond the symptoms and morbid tissue changes
to the disturbance of the normal relations that cells,
or aggregate of cells, or the organism, bear to their
environment. It must not be directed solely against
the symptom group, nor be controlled by the morbid anatomy alone, nor find its only guide in the
perversion of the physiological processes. The
chronic disorders of digestion and nutrition are, in
their incipiency, the expression of either a chemical
lesion of the fluid in which the cell lives, or a nutritive defect in the structure and composition of the
cellular protoplasm. It is the alterations of the
composition of the fluids of the body, and the consequent indefinable defect in the structure of the
cellular protoplasm so intimately related to the un-

healthy variations in the functions of the cell, which
make up the canvas on which the clinical picture is
painted in the colors of morbid anatomy. It is the
faulty assimilation of imperfectly prepared nutrient
material, as embodied in a badly constituted protoplasm, that gives the progressive quality to these disorders, and the one hope of relief and the grand purpose of treatment is to secure in some way a better
quality of cellular contents, to correct this nutritive
defect. This cannot be done by starvation any more
than it can be accomplished by forced feeding. It
is the combination of means that will remove as
well as build up, that break down as well as regene-
rate, which will yield a permanent result. An extreme nicety of the digestion and of the preparation
of the nutrient material, its proper distribution to
the tissues by the blood, the quick solution and removal of waste products, and the conservation of
nerve force, place the cell in an environment which
is most favorable to its nutrition and life. The
promotion of a high degree of healthy nutrition is
the one essential purpose ; and, in the present state
of knowledge, cell life can be appreciably influenced
or controlled only by modifying its environment.
A treatment based on this principle is causative and
curative, and scorns a plan that only aims to secure
the suppression of symptoms.

As we have already seen, these chronic disorders
arise from the persistent and almost imperceptible
disturbance of the continuous adjustment of relations as manifested in the life processes—the integration of structure, the evolution of force, or the
elimination of waste products. The diatheses are

examples of these evil tendencies indelibly stamped
upon the organism in the process of its making,
and may manifest themselves directly or indirectly,
through nerve or blood or lymph changes, as disorders of the digestive system and its appended
glands. Now it is a predisposition to the development of lowly organized cells—a defect of nutrition
in which the power of assimilation is in abeyance
—a vice of constitution in which the tissues yield
readily to incident disturbance and have little con-
structive power ; of such a nature is the inherited
nutritive dyscrasia which forms so favorable a soil
for tuberculosis. Now it is the fibrous tissue which
shows the evil tendency and stamps the organism
with the fibroid diathesis. Or it is a hæmic or hepatic state that manifests itself as rheumatism or
gout. Now it is a fault of the more highly evolved
nerve centres, and the patient falls a victim of some
neural disorder or is skilfully conducted through
life on a sleeping volcano. Or it may be some defect of elimination which permits the accumulation
of some such waste product as uric acid in the system. These evil tendencies, when inherited, which
underlie many cases of disordered digestion, cannot be completely eradicated by treatment, and our
purpose in therapeutics is limited to the prevention
or control of the manifestations.

Again, the treatment of these chronic disorders
is limited by destruction, degeneration, and atrophy of the anatomical elements, and by deformity.

In acute disease the incident disturbance falls
directly on the functionating cells, which recover or
redevelop more or less completely when the morbid

influence passes away; or function is perverted by the compression of unorganized inflammatory products. In chronic disease the cells are sometimes involved indirectly by the formation of new tissue and by compression. The new connective tissue may simply irritate, but it usually contracts from age and dies itself and destroys other neighboring tissue. A cure is possible only in the formative stage, when the chronic productive inflammation may resolve. Therapeutics is thus limited by the nature, relations, and age of the pathologically formed tissue.

Chronic disorders, again, often arise from degeneration or atrophy or deformity. The persistence of the symptoms is due to the persistence of the damage done by former disease. We can do nothing in a medical way to remove the cicatricial stenosis of the pylorus or stricture of the bowel. When chronic disease falls directly on the anatomical elements of an organ, it is commonly a degenerative or atrophic process, and if the cells be reproduced they are imperfectly and lowly organized. When gastric atrophy occurs from age, little can be done to stay the progress of decay, for it is usually accompanied by a like condition of the duodenum and of the other viscera; life slowly dissolves beneath its burning rays. But when atrophy occurs in the developmental or vigorous periods of life, as in the gastric atrophy following typhoid fever, or the intestinal atrophy resulting from prolonged distention by the gases of organic fermentation, treatment is limited but of some avail. An accurate anatomical diagnosis defines and limits therapeutics.

Having briefly indicated the primitive nature and the basis of the cure of these chronic disorders, and advocated the necessity and the utility of a well-defined and comprehensive treatment, turn we to a consideration of the remedies to be systematically employed.

Our therapeutic purpose in chronic disease is never so narrow as the prescription of this or that drug; it is the combination of many means to meet complex indications, the treatment of the whole man as disturbed by disease. As we grow old and gray in the service of our calling, the less do we rely on drugs alone. By the proper use of drugs we can often snap the thin-spun thread of evil sequences, and we will not be persuaded to cast away means of such power and precision. I believe our object is best accomplished by a systematic combination of remedies. And our first aim should be the promotion of a high degree of healthy nutrition with a view to increasing the resistance and activity of the tissues and to securing physiological cell structure; and, secondly, the regulation of the patient's life and diet with a view to the readjustment of the damaged organism to vital demands; and, in the third place, the rational use of drugs as based on their physiological actions and as confirmed by clinical experience. This forms the great tripod of treatment.

If one will take the trouble to turn through medical literature he will be surprised to learn the conspicuous part which has always been assigned in ætiology to "impairment of the general health." In many cases of acute disease the most robust consti-

tution yields to the shock of the violent onset. It
is, however, more often the weak and tired who are
forced to the wall. But a well-nourished body not
only resists invasion ; it also limits and conditions
and controls the morbid process—has a curative
power. A problem to solve in all of these chronic
disorders is the problem of nutrition, and upon its
solution depends the possibility of relief. And it is
not enough to adapt the quantity and quality of the
food to the vice of nutrition we wish to correct or
the state of nutrition we wish to establish, though
this is of very great importance. It is not enough
to adapt the quantity and quality of the food to the
present state of nutrition, the capability of the digestive organs, the activity of the emunctories, and
the evolution of force as conditioned by habits of
life and environment—though if this be not done
success will rarely crown our efforts. But the patient must be kept under daily supervision, and the
physician must see that the diet is fulfilling its therapeutic purpose, and readjustments be made to meet
the varying indications afforded by the clinical
guides to nutrition and digestion. The ability to
use one's knowledge in the treatment of disease is a
distinguishing mark of the practical physician. In
the chronic disorders of the digestive tube it is essential to have it made clear to us how the food is
being worked up and utilized in each particular
case. This cannot be easily determined with exactness ; our guides are not absolute, because our
knowledge is not complete. But this is no reason
why we should not employ them so far as we know
them worthy of trust. I could just as willingly and

easily dispense with physical examination in the diagnosis and treatment of the diseases of the heart
and lungs, just as well omit the microscopic and
chemical examination of the urine in the diseases of
the kidneys, as to turn out the light thrown on the
disorders of digestion and nutrition by the examination of the stools, the urine, and the blood. We
know the qualities and number of the corpuscles of
healthy blood, and we know the percentage of hæmoglobin it contains ; and by the microscope, hæmocytometer, and hæmoglobinometer we can tell
when a particular unhealthy variation is approaching or falling away from the normal standard—this
is the index of assimilation. In a similar manner
the urine gives a good deal of testimony concerning
disassimilation, hæmolysis, and digestion. And
something about the condition of the digestive tube
may be read in the stools. When the information
obtained by these persistent examinations is supplemented by the knowledge gleaned from the subjective symptoms and the physical signs, I have not
found it very difficult to arrive at pretty accurate
conclusions concerning the state and efficiency of
digestion and nutrition. It is one thing to calmly
preach, from an office chair, diet theories to a patient struggling for relief ; it is another to stand
sympathetically by his side and see that he digests
and assimilates what he eats. The diet must be
prescribed on comprehensive, scientific principles,
with a clear, well-defined object in view, and be
brought to the test at the bedside ; for the clinical
test is supreme, and educated common sense must

count for something in the dietetic management of the chronic disorders of the digestive tube.

It is not enough to send the cells a fluid rich in oxygen and in nutrient material, for it must also be free from poisonous products. The circulation of a pure and rich lymph must be active, so that there be no accumulation of cellular waste. Active oxidation is a strong barrier against auto-infection if the poison succeed in passing through the liver. But the one great remedy for auto-infection is free elimination by the kidneys, intestine, lungs, and skin. The best solvent, the best diluent, and the best diuretic is a plentiful supply of fluid ; and to liquefy the bile and promote its discharge, to excite normal peristalsis and cleanse the alimentary canal, this fluid should be taken hot. In the Cavendish Lecture of 1891 Dr. T. Lauder Brunton speaks of the great value of hot water in the treatment of gout, rheumatism, and lithaemia. It is the most efficient and the safest eliminating remedy that we possess. The use of cholagogues, diuretics, and diaphoretics will prove of some value. To secure a high degree of healthy nutrition we need a rich, actively circulating, and pure lymph. Every cell must have a clean lymph in which to bathe and from which to draw its life and strength.

Second in importance only to careful alimentation and active elimination are the control of the habits of life and the selection of favorable surroundings. Here the indications are so special in individual cases that little can be said in a general way, and much must be left to the physician's common sense and experience. Under this heading must be included many remedies the value of which is well recognized—a favorable climate, pure air and sunshine, bathing, exercise, rest, massage, electricity, a contented and hopeful mental state, etc.—all contributing to the ease or activity of the circulation, of respiration, and, briefly, of all the secretory, excretory, and nutritive processes. Physiological living is a great remedial power, and we should never grow weary in the enforcement of healthy physical, mental, and moral habits. Disease will not become less so long as the people through ignorance do not take care of life. Enough vitality is destroyed in riotous living, morbid thinking, and useless and often causeless worry to add to human life another score of years. And it is particularly this large class of neurotics and dyspeptics who do not know how to live economically and conserve energy. The physician must teach them to do so before he can hope to cure them.

The tendency of modern drug treatment is local and special. In addition to local antiseptics, we have also drugs that tend to keep the fluids of the system sweet and that affect one or more of the functions of an organ in a special way. It may be hoped that we may some day be able to directly endow special cells with particular powers, and the fond hope is not without some foundation. We have no so thoroughly efficient local treatment that it cannot obtain some help from constitutional measures, and we should not forget the remedies that aid nutrition, regulate elimination, and control neuro-muscular discharges. Drugs that relieve gross symptoms are also of very great temporary value. Pain is in itself an evil and must be quieted. Nervousness must be controlled. Insomnia requires a hypnotic. But to the aid of drugs must be brought other and more powerful means—the promotion of a high degree of healthy nutrition, free elimination, well-ordered habits of life, and a favorable environment. Here lie the hottest of the battle and the hope of victory.

In the management of chronic disease tact and common sense are worth almost as much as medical knowledge. The course is a long one and tests the endurance of the physician. Such is the solidarity and such are the intimate relations of the nutritive processes that an unhealthy variation of one soon forces the others to fall into harmony with it; consequently these disorders are not self-limited, but progressive. And it requires as much time to re-establish the normal state as to arrest and correct the primitive perversion of force. I would emphasize the importance of long-continued supervision and minute instructions. The physician, as does the surgeon, succeeds most often when he is a strict observer of detail, when he knows and remembers and does little things.

My plea is for a broad and comprehensive and well-defined therapeutics ; a plea for the paramount importance of hygiene and dietetics ; a plea for the considerate use of drugs ; a plea for the bedside study of this highest comportment of medical knowledge in which science and art lie down together.

CHAPTER II

THE CHRONIC DISORDERS OF GASTRIC DIGESTION.

THE clinical therapeutics of the diseases of the stomach is a subject of great practical importance. The diseases of no other organ come more frequently under the care of the physician, produce more annoyance or suffering, and yield more surely to judicious treatment.

This chapter on the chronic disorders of gastric digestion is not intended to be an exhaustive one. Ætiology, pathology, and symptomatology will be considered only in so far as they bear on differential diagnosis and treatment. The cure of any chronic disease is largely comprised in its ætiology, and a correct diagnosis is an essential preliminary to rational treatment.

It is not often possible to make a complete anatomical diagnosis of a disease of the stomach. Moreover, morbid anatomy is only a symptom, and a lesion of the mucous membrane is not always present. Neither is an ætiological classification practical. The same cause may originate a variety of disorders. Alcohol may produce hydrochloric superacidity or subacidity, or gastritis. Tuberculosis may be accompanied in its early stage by chemical or motor insufficiency of the stomach, with cough and vomiting from the irritation of the supersensitive ends of the vagus by the food. The inadequacy is not due to a gastric lesion, but to tubercular toxaemia. Hydrochloric superacidity sometimes aids in the preparation of the nutritive soil. Gastritis with a raw and fiery tongue, anorexia, and diarrhœa is the form which belongs to advanced phthisis. Chronic digestive disorder, with and without a lesion of the mucous membrane, seems to be the most useful general classification.

However closely dyspepsia, in the end, may be associated with errors in diet, the derangement of the process of digestion is nearly always due, in the beginning, to disturbance of cell secretion or to impaired muscular movements. There is no lesion of the mucous membrane. Hence dyspepsia may be briefly defined as gastric insufficiency without alteration of structure.

The impaired movements and defective secretion are the local manifestations of a constitutional state. Who would find the cause of dyspepsia must look beyond the stomach to the thin and impure blood, to the weak and tired nerve centres, to impaired cell activity throughout the body. Perverted secretion is often the result of defective cell nutrition. The fault may lie in the lack of tissue-forming material in the blood ; or this important nutritive fluid may be surcharged with the products of defective metabolism, or with poisonous material absorbed from the alimentary canal or left in the circulation in hepatic or renal insufficiency. Thus we find it in the anæmias, chlorosis, gout, chronic rheumatism, lithæmia, malaria, syphilis, and chronic nephritis ; or it may prove to

be the legacy of former acute illness or infectious
disease. Tuberculosis and alcoholism are also common causes. But the chief factor in the causation of
dyspepsia—always present, always active, affecting
either secretion or muscular movement, or both—is
impaired nerve supply. This weakness or pervers-
sion of the regulating or controlling action of the
nervous system may be of central origin or reflected from a distant or functionally associated organ. The great clinical masters have often noted
the frequency with which dyspepsia occurs in the
neurotic—an individual with congenital instability
of nerve. The part that heredity plays preponde-
rates ; but impaired innervation is not rarely the
result of the reckless perseverance and unrest of
modern life. Dyspepsia finds many a victim on the
rugged highway along which honors lie to be gathered and worn. Sudden reverses of fortune, intense emotion, moral shock, great sorrows, the
prolonged strain and often intense agony of the
critical periods of life, leave exhausted nerves and
dyspepsia. These patients are all primarily, or as
a result, neuropathic. Nothing further need be
written, we hope, to impress the principle that, if
we wish to cure dyspepsia, our therapeutic purpose
must reach beyond the stomach to the underlying
defect of constitution, or vice of nutrition, or pathological nerve state.

But this is not all. Defective alimentation—overeating, improper food, and, indirectly, starvation
through exhaustion of the nerves of organic life—
must be considered as a possible cause by destroying
the equilibrium that obtains in health between the

quantity of gastric juice secreted and the chemical work required of it. Thus the stomach is unequal to its task. An excessive use of the carbohydrates is a well-recognized cause of lithaemia and of neurasthenia, and these originate dyspepsia through their depressing psychical states and the exhaustion of the nerves presiding over secretion and muscular action. There is, however, no satisfactory reason for doubting that errors of diet always give rise, at all events in the beginning, to the lesions of gastric catarrh.

Chronic gastritis is frequently the sequel of dyspepsia. It often follows the acute disease, which is only rare because it is not recognized. Occasional transgression of dietetic laws seldom results in persistent pathological changes. Habitually recurring patchy congestion, initiated by some mechanical or chemical irritant contained or developed in the food, may not subside while the stomach is taking its rest, and the tissues and nutritive processes are moulded into conformity with the morbid condition. Passive congestion of the stomach in disease of the liver, lungs, heart, or spleen, or from venous obstruction by the pressure of a tumor or enlarged gland, is accompanied by the secretion of a large quantity of mucus and a diminution of hydrochloric acid, and undue fermentation with its consequences results. Anaemia with its weak heart acts in a similar manner. Chronic gastritis may be secondary to renal disease, or form a part of the history of arterio-capillary fibrosis. Not a few of the most obstinate cases have for their cause an endarteritis of long standing, or amyloid degeneration from exhausting purulent formation. Again, a naso-pharyngeal catarrh or bronchitis may initiate and feed the fermentative process by the decomposing mucus or pus finding its way into the stomach. The successful management of chronic gastritis depends largely on the detection and removal of the underlying cause that gives its type to the disease.

Gastric ulcer is a common lesion. It is pre-eminently a disease of women, and is usually preceded and accompanied by a disease of the blood, such as anaemia, chlorosis, or the thin blood after labor ; by diminished alkalinity of the blood, as in oxaluria and uricaemia ; and by hydrochloric superacidity of the gastric juice. Virchow's theory that the simple ulceration results from the plugging of the nutrient artery of the part and digestion of the infarct, seems to satisfactorily explain many cases. The endarteritis, thrombosis, or embolism may be the consequence of the diminished alkalinity and the fibraemic state of the blood. Syphilis and tuberculosis are well-known causes of the specific ulcers.

A predisposition to cancer is inherited ; its development is supposed to be excited by chronic irritation, and, probably, by a specific germ.

Atrophy of the gastric glands is due to parenchymatous or interstitial inflammation, to acute or chronic degeneration, to infectious or wasting disease like typhoid fever or tuberculosis, and to innutrition from prolonged distention or extreme dilatation of the stomach. Chronic inflammation, ulceration, cancer, and atrophy are the lesions of

2

the mucous membrane which often accompany the chronic disorders of the digestive process. But gastric inadequacy is not always manifested in morbid tissue changes. What are the varieties of dyspepsia, and in what way can we detect the unhealthy variations from the physiological process?

The modern methods of examining the gastric juice are familiar to the profession and need not be reviewed in this short chapter. Always of value, the analysis is in some cases essential to a correct secretory diagnosis, and often enables us to see where we could only guess at the truth. It is a useful guide in the administration of drugs to supplement the gastric juice. But it is too great a burden to the physician and too disagreeable to the patient to become popular with the profession. Stomachal chemistry is of very great scientific interest, is an aid to treatment and diagnosis, but it is not so easy, nor so essential, nor so clear in its suggestions (the inferences drawn from it are remarkably contradictory) as to allow one to conscientiously urge its general adoption. A word of warning should also here be given. Stomachal chemistry is reducing treatment to a very simple formula--hydrochloric superacidity demands alkalies in large doses, subacidity indicates the administration of hydrochloric acid. We shall see, a little further on, how narrow and irrational is this method of treatment. Again, it is assumed that a certain secretory defect is so indelibly stamped on the mucous membrane that it continuously goes wrong in this way and in no other. This is a general truth applicable to the grand types. But no other organ is

so fantastic and variable in its work as the stomach—a thought, a feeling, an emotion may influence it ; and to a degree its secretion changes with
every varying stimulus or nerve state or blood
supply.

The most important constituent of the gastric
juice, from a pathological point of view, is the
hydrochloric acid. It is not the state (combined or
free) of the hydrochloric acid, but the quantity secreted by the mucous membrane, that is the guide.
A large quantity of albumin requires only a very
small amount of pepsin for its hydration in a proper
medium ; and it has not been demonstrated that too
little pepsinogen is ever secreted in any other condition than glandular atrophy. Theoretically this is
true, but practically the administration of pepsin is
of great utility when it is necessary to prescribe a
largely nitrogenous diet in glandular atrophy, combined with hydrochloric acid, and without hydrochloric acid in defective absorption. The presence
of a large quantity of peptones arrests peptonization, but the process of hydration recommences on
the addition of a new supply of pepsin. The abnormal quantity of hydrochloric acid secreted is the
index of the disturbance of the second and preparatory stage of the successive development of the digestive process, which reaches its climax of chemical changes in the intestine. But from a clinical
point of view gastric motility is even more important than gastric secretion. When the movements
of the stomach are perfect and the pylorus does its
work efficiently, there are no gastric symptoms unless the mucous membrane be supersensitive. The

stomach does a grand chemical and preparatory work of its own in peptonization, uncovering starch, liberating fat, and unbinding muscular tissue ; but it is its duty also to protect the duodenum and to dispense to it slowly and within the right time the properly prepared and well-mixed chyme. It is on the unhealthy variations of hydrochloric acid and the abnormal muscular movements of the stomach that we have found it of most value at the bedside to base the classification of dyspepsia, and it is accordingly as these two factors are increased, diminished, or irregular that a deviation from the state of health can be said to exist.

Gastric dyspepsia with increased formation of hydrochloric acid is usually associated with one of the neuroses, and occurs in two varieties—superacidity and supersecretion. In simple superacidity the fasting stomach is found empty ; in continuous supersecretion the stomach in the early morning, before eating or drinking, contains one or more ounces of gastric juice, which may or may not be superacid. Both predispose to gastric ulcer (markedly so in anaemic women), and are frequently accompanied by dilatation from pyloric spasm and organic fermentation, and sometimes also by downward displacement of the stomach and of the first part of the duodenum. The hydrochloric acid may be secreted in such quantity and so rapidly as to at once stop the action of the saliva, which should continue in the stomach from ten minutes (Ewald) to half an hour (Van den Velden). Organic fermentation does not occur unless—as may happen when dilatation and more or less atrophy are present—the hydrochloric acidity falls below 0.7 per cent. This is theoretically true, as may be demonstrated in a test tube; but, clinically, in the stomach we not infrequently find organic ferments and fermentation when the gastric juice is superacid from excess of hydrochloric acid. In simple superacidity the appetite is increased; eructations are extremely acid, but usually without much gas; epigastric pain is paroxysmal and severe, comes on soon after meals, and is often relieved by the ingestion of water and nitrogenous food. Proteids and albuminoids are rapidly digested when the food mass is small and permeable, the fats are partly decomposed by the free hydrochloric acid and the fatty acids give rise to heartburn, and intestinal digestion is delayed or arrested by the superacidity of the chyme. Diarrhœa is often present. The stomach wall is tonically contracted with painful peristaltic waves. The urine is quite alkaline during digestion, but regains its normal acidity, or may become excessively acid, in the interval. In supersecretion the appetite is variable; eructations are very acid, often fetid and gaseous; pain is more or less continuous, becoming paroxysmal (immediately or) about three hours after meals and about 3 o'clock in the morning, and is almost completely relieved by vomiting. Gastric digestion is slow and imperfect, and gastric absorption is very much diminished. The vomit is sour, often foul and of acetic odor, and contains organic ferments and undigested food eaten a day or two before. The signs of dilatation are present, the greater curvature is on a level with or below the umbilicus, and morning splashing and sometimes seething are easily elicited. The stomach may be distended and the pylorus displaced downward, and small quantities of bile frequently regurgitate, or in other cases (which are somewhat rare) almost continuously flow, into the stomach. Constipation is the rule, but morning diarrhœa is not rare. The urine is almost continuously alkaline and precipitates the phosphates.

These varieties seem to be stages (the duration of which is very variable) in the orderly development of one disease. What is the probable explanation? The digestive disturbance seems to begin with supersensitiveness of the nerves of the mucous membrane and consequent excessive tonicity and excessive (and sometimes continuous) secretion through over-excitation of the motor, vasodilator, and secretory nerves. Dilatation here does not result from atonicity of the muscular coat. Its pathogenesis is the same as the dilatation above an intestinal stricture, the same as the dilatation of the left ventricle in aortic stenosis, the same (and the analogy is very close) as the dilatation of the bladder from excessive irritability of its neck or of the deep urethra. The pylorus, the powerful automatic protector of the duodenum, contracts and obstructs; the gastric walls become irritable and hypertrophy; fermentation and distention and loss of compensation and dilatation supervene. Diminished absorption, mucous catarrh, destruction or atrophy of the gastric glands, and diminished secretion complete the picture. Ulceration may occur at any stage. On this interpretation will be based the curative treatment, so far as it depends on the administration of drugs.

Dyspepsia with diminished formation of hydrochloric acid is met with most frequently in individuals with "weak stomachs." Digestion is slow and the lactic-acid stage is prolonged. The excess of lactic acid is formed from the sugars (and in small quantity from the starches) through the agency of numerous fermentation organisms; it may be split up into water, butyric and carbonic acids, and hydrogen. The hydrochloric acidity, even at the height of digestion, does not often rise above 0.7 per cent. A little too much work or mental worry and a little too much food suffice to derange the digestive process. The flatulence and acidity are most marked two or three hours after meals.

Irregularity in the secretion and muscular movements of the stomach is due to sympathetic disturbance. The stomach, through its complex nerve connections, is in intimate relation with nearly every organ in the body. Habitual speedy vomiting without preceding nausea is nearly always reflected. The gastric disturbance comes on suddenly and without warning, and varies in kind from day to day. In individuals with impressible nerve centres and weak inhibition the stomach is the organ toward which every little local storm seems to wend its way.

Exaggerated muscular movement of the stomach is a rare derangement of the process by which food is made ready for assimilation. The pathological unrest falls primarily on the muscular layer. The exaggerated peristalsis commonly extends to the intestine, the two being intimately associated in their movements—a principle utilized to excite a free discharge of bile and to cause a dilated stomach to empty itself by means of the cold-water enema, and not only the solution but the absorption of aliment is prevented. It is often associated with hyperaes-thesia of the mucosa and a ravenous appetite and obstinate insomnia.

Gastric atony incidental to a state of weakness and relaxation of the whole muscular system is a common gastric defect. Brain workers who lead sedentary lives furnish the largest number of its victims. The muscular layer lacks tone and peri-staltic movement is weak. The face wears an ex-pression of fatigue ; the heart is weak and irritable, and arterial tension is low : the muscles of the throat are flaccid—there is a general want of tone. The gastric juice is normal : digestion is slow, but complete if not interfered with by fermentation. The appetite is unimpaired and the bowels are con-stipated. A sensation of uneasiness rather than of distinct pain : a feeling of weight or heaviness from long-continued pressure of the food on the same spot : flatulence from muscular weakness and vaso-motor relaxation (as in the intestinal paresis of peri-tonitis), or from regurgitation of gas through the open pylorus—complete the clinical picture. The extreme nerve-tire explains the want of muscular tonicity, and the weak stomach, on account of the prolongation of its labor, gets little rest. Unless the process be controlled by judicious treatment and the organ and system strengthened, extreme dila-tation will surely supervene. These patients are only cured by the combined treatment of digestion, nutrition, and the nervous system. It is a profound error to throw them into the great drag-net—neurasthenia.

There is another form of gastric atony that frequently comes under the care of the physician, and which dates its beginning in early life. One cannot closely study these cases without detecting heredity's powerful hand in their development—a variety of the "weak stomach" in which the inherited or early acquired or early manifested defect falls on the muscular rather than the secretory system. The muscular layer is undeveloped, atrophic as well as atonic, and peristaltically weak. Atrophy of the gastric and intestinal glands may rapidly follow dilatation, and death from malnutrition close the scene before the morning of life has passed ; or the curse may be suspended while the years roll by, until finally the sword falls and " slits the thin-spun life." The stomach may be strengthened by careful feeding, but the vice of constitution is irremediable.

Dilatation in either form of gastric atony is commonly associated with a like condition of the large (and small) intestine. Malnutrition of the local ganglia in all probability has something to do with the glandular atrophy. In no other condition do the symptoms of auto-infection become so prominent. The epithelium throughout the alimentary canal is lowly organized, and here and there the wall is as thin as parchment and free from glands. These pouches (favorite sites of which are the cæcum and hepatic and splenic flexures of the colon) are filled with decomposing and fermenting fæces. The peptones fail to be reconverted into serum albumin, and the emulsified and split-up fats cannot be built up into glycerin neutral fats on their way through the mucous membrane to the central lacteal or blood vessel. Absorption is imperfect, unselective ; assimilation is disordered and nutrition fails. Emaciation is marked, and the products of fermentation and decomposition and incomplete digestion absorbed from the alimentary canal congest the liver, irritate the nerve centres, and inflame the kidneys. Hysteria, insomnia, or a demon-like melancholy which no effort can throw off fastens itself on the victim. The clinical history, self-infection, the absence of hypertrophied walls and visible movements, easily exclude dilatation from pyloric obstruction, functional and organic.

Three methods have been suggested, apart from the clinical history, subjective symptoms, and physical signs, to aid in the diagnosis of motor insufficiency. A pint of olive oil (Klemperer) is introduced, and what remains of it in the stomach after two hours withdrawn. The difference represents the quantity that has gone into the duodenum. The objections to this method, and the liability to errors, are too great to allow serious consideration. Of more utility is the salol test, if the kidneys are sound and the reaction in the duodenum is alkaline. In health salicyluric acid appears in the urine in half an hour (Ewald and Sievers), and disappears in twenty-four hours in health, thirty-six hours in atony, and forty-eight hours in dilatation (Silberstein). The third and a very good method is to

administer Leube's test meal or Ewald's test breakfast, and examine the contents of the stomach at
varying periods thereafter.

The differential diagnosis of dyspepsia and chronic
gastritis requires close study and careful reasoning.
The two diseases merge into one another, and in
vague cases without clear-cut features it is difficult
to learn at the bedside with which form we have to
deal. The history of the case, the order of appearance, and duration of the symptoms must be taken
into consideration in the formation of a conclusion.
The known nature of the disease to which the gastric disorder is secondary may help to clear up the
obscurity. A careful chemical or microscopical
examination of the blood, of the gastric juice, and
of the excretions will always prove of value.

The local signs of chronic gastritis are persistent,
while those of dyspepsia are intermittent and ca-
pricious. The pain of chronic gastritis is more
severe when the stomach is full ; in dyspepsia it
may occur only when the stomach is empty, and be
relieved by taking food. Violent paroxysms of pain
in chronic gastritis are made worse by pressure ; in
dyspepsia firm pressure may give relief, and an
interval of comfort follows each attack. Repeated
vomiting of mucus, or of mucus mixed with undigested food, is pathognomonic of catarrhal gastritis.
Increased hydrochloric-acid formation is present
only in dyspepsia ; in chronic gastritis the quantity
of hydrochloric acid is diminished. This is true as
a general rule. In the beginning of gastritis the
irritable mucous membrane not rarely supersecretes
a superacid fluid ; this is particularly true if the

gastritis be the sequel of a secretory neurosis. Thirst, nausea, and anorexia are more frequently linked to an alteration of structure. In certain mild forms of gastric catarrh a morbid sensation, closely allied to the sense of hunger and radiating backward between the scapulæ, recurs at regular intervals; its disappearance on the taking of the first mouthful of food is followed by nausea; the slight irritation of the food seems sufficient to produce dilatation and stasis of the blood current in the previously hyperæmic mucosa. In simple catarrhal gastritis there is excessive secretion of mucus. The symptoms vary very much with the extent and destructiveness of the inflammatory process, with the degree of glandular atrophy and dilatation. The dilatation is mechanically produced by the accumulating mass of fermenting food, or by infiltration of the muscular layer by inflammatory products; when well marked its diagnosis presents no difficulty. It is not on any one sign, but on the symptom group and the results of the examination of the contents of the stomach, that the diagnosis must be based.

The diagnosis of gastric atrophy can be based with certainty only on the long-continued absence of hydrochloric acid, pepsin, and the lab-ferment, as proved by repeated examination of the gastric juice. If there be no stasis of the food mass in the stomach, the duodenum may completely supplement the gastric insufficiency and no symptoms of dyspepsia make their appearance.

The symptoms of ulceration are those of superacid dyspepsia—local pain, hæmatemesis, and local tenderness. The tender spot is usually circumscribed and located a few inches below (and to the left of) the tip of the ensiform cartilage ; its diagnostic features are its strict localization and its persistency. The pain of the accompanying superacidity is relieved temporarily by food and drink ; the pain of the ulceration is increased or excited by eating. Hæmorrhage, if it occurs, is usually profuse : care must be taken to exclude acute inflammation, portal obstruction, cancer, and toxæmia. Enlargement of the spleen is also accompanied by gastric hæmorrhage. In hepatic cirrhosis the blood may come from rupture of a dilated œsophageal vein.

Ulceration may occur at any age, is of indefinite duration, irregular in its progress, and is often relieved and cured by treatment. Even with the aid of a complete clinical history, of the subjective symptoms and the physical signs, we may be unable to state whether ulceration is or is not present. Perforation may be the first and only sign. Sudden and large intestinal hæmorrhage (large, tarry movements), preceded by paroxysmal gastralgia, extreme pain in the right hypochondrium. and more or less duodenal dyspepsia of long standing, are the symptoms of duodenal ulcer. In gastric ulcer the constitutional state is proportionate to the digestive disorder.

In cancer the patient is above 40 years of age : hæmorrhage is small and slow ; there is rapid and progressive decline, and cachexia ; hydrochloric acid soon permanently disappears from the gastric juice : a tumor may be felt; treatment gives little relief,

and the constitutional state is out of all proportion
to the disturbance of gastric digestion.

The exact diagnosis of disease has its peculiar
charms; at all events, in difficult cases it is the
flowering of medical science. But after the flow-
ers should come the fruit. Turn we now to treatment—to the consideration of the moral management, hygiene, diet, and medicinal agencies which
clinical experience has shown to be of value in the
palliation or cure of the chronic disorders of gastric
digestion.

The moral management of these diseases has not
received the attention that it merits. We wish to
urge its importance in the cure of those cases in
which the weakness or derangement of the central
nervous system is well marked—when this state is
a primary ætiological factor. In the cure of neurasthenic dyspepsia it is the keystone to the arch;
it is the one means of rolling away the cloud that
darkens the pathway of the neurotic. These individuals have no will power or reserve force, and in
no other way can we aid them in throwing off the
delusion that they are incurable. It is our duty to
make every endeavor to impress the patient with
the fact that his case is thoroughly understood. A
correct anatomical and pathological diagnosis will
enable the physician to state with precision what
can be done. Firmness and kindness of heart are
the means of winning confidence. Faith, inspired
by truth, honesty, and manly bearing, stimulates
and tones the nervous system and unbinds the will.
No one doubts the power of expectant attention.
Digestion is dominated by the nervous system, and

the centres controlling secretion and muscular movement are re-represented in the cortex. The physi
cian who fails in the moral management loses an
essential aid in the cure of these chronic cases.

Not the moral management alone is of importance ; the life of the patient must be on a physiological basis. Insist on slow and regular eating,
and not too great a variety. The stomach is only
confused and disordered by course dinners. A rest
of half an hour before and an hour after each meal
is a duty. Clothing should receive consideration,
and in our climate the whole abdomen should be
protected at all times by a knitted bandage of wool,
wool and silk, or silk. The elasticity supports also
the dilated stomach and gives comfort in obesity.
The method and frequency of bathing should be
suited to the patient's general condition. Careful
attention to every detail is the price of success.

Hours of work, recreation, and rest are to be pro-
portioned to the severity of the case. In the mild
cases the patient should live in the open air during
the hours of sunshine. A daily drive, or a ramble
and view of a favorite landscape, may lift the mind
away from self and the worries of business and life's
daily cares. In the severer cases confinement in-
doors may be obligatory, the bedroom must be
kept full of fresh air, and the day be spent in quiet
enjoyment in a sunny room. In the grave cases,
when the nervous system is a wreck and the function of every organ in the body is in abeyance—a
condition closely allied to prolonged shock—isola-
tion, absolute rest in bed, massage, electricity, oxygen inhalations, and a tissue-building diet will fre-

quently enable the patient to emerge from the restorative process fresh as if from Medea's charms.

But of more importance than all else in the treatment of these diseases is the selection of a proper diet. This is "the great and master thing"- the question of feeding. And right here it is essential that we should clearly define the principles which may best guide us in the adaptation of a diet to individual cases of disease.

And first we must protest against the guidance of a morbid appetite and of morbid desires. The "natural instincts" of the patient must not "have free play," though "they have grown up under the regulating force of universally acting biological laws, under the pressure of the sleepless vigilance of the law of survival of the fittest, and the sure incidence of the laws of heredity" (Sir William Roberts). It might be well to suggest the possibility of the development of types from unhealthy variations, which might serve fittingly to illustrate the self-avenging power of Nature's laws. Every form of force is modified by the nature of the medium which manifests it, and the "natural instincts" of the invalid are no better guides to alimentation in disease than are the delusions of insanity guides to conduct.

"Find out that course of life which is best," writes Pythagoras, "and habit will render it most delightful." If reason, then, must define the diet, on what knowledge should its dictates be based? The answer is a simple one—on "the rational standard of diet, as revealed in the customs and habits of the people," as Sir William Roberts rightly ob-

serves, and as corrected by the known digestibility
and nutritive value of the various articles and
classes of food; on the capability of the digestive
organs; and on the state and needs of general nutrition.

A cursory view reveals the fact that the inhabi-
tants of the temperate zone live on a mixed diet of
proteids, albuminoids, fats, and carbohydrates. It
would be interesting to know something of the
effect of these classes of food on destructive metabolism and the building-up of tissue. The albuminoids and proteids increase nitrogenous waste.
When administered along with the fats or carbohydrates in sufficient quantity to supplement and
raise the force evolved in the splitting-up of the albumin in the circulating fluids to the level of the
requirements of the vital processes, or when the
storage of fat in the system can be utilized for this
purpose, none of the cells of the body are destroyed.
When the quantity of albumin circulating with the
nutritive fluids is not all required to meet the demands of the vital processes, within certain limits,
as defined by the inherent activity of the cells and
that delegated or withheld by the nervous system,
new cells are generated. Peptones furnish energy,
but do not form tissue. Tissue is built up out of
unchanged or incompletely digested proteids and albuminoids. Thus albumin is the great sustainer of
life, and, under proper conditions, the great builder
of tissue. It cannot be supplanted, beyond a certain
point, by any other food. It makes the blood richer
in red corpuscles and in hæmoglobin, as any one
can easily demonstrate by the hæmocytometer and

3

haemoglobinometer in anaemic and chlorotic dyspeptics on an exclusively animal diet. It is the only class of food that can alone support life, and it forms the physical basis of life in its simplest and primordial form.

The assimilation of the fats is aided by the proteids and albuminoids. Fat diminishes nitrogenous waste, and is intimately concerned in the nutrition of the nervous system, and forms nearly all the fatty tissue of the body. The carbohydrates never enter into the formation of tissue, but aid the organization of albumin and fat by supplanting them in destructive metabolism. Thus it is evident that the nutrition of the body can be most economically maintained at a high point by a due admixture of these three classes of food. But in disease the capability of the digestive organs, or, in the special diseases under consideration, the capability of the stomach, imperatively demands a compromise. But an early and cautious return to a suitable mixed diet will suggest itself to the common sense of the physician as the best method of avoiding the evils of exclusiveness. The excessive or exclusive use of the carbohydrates tends to dilatation and disease of the stomach and intestines, and the individual is pale, thin-blooded, weak, and bloated. A long and exclusive use of the proteids and albuminoids tends to certain circulatory derangements and to nervous irritability; while the malassimilation of fats is the most important factor in the production of the emaciation in the pre-bacillary stage of tuberculosis. The physician, like the general he should be, must avail himself of every opportunity to advance, or be ready to retreat under cover on the first note of warning, until his object has been attained.

In the meantime the strength of the enemy must be correctly estimated, or, to drop the metaphor, the capability of the stomach must not be exceeded. That all of the food taken undergoes digestion and absorption is made known by the absence of the clinical signs of fermentation or putrefaction, but chiefly by the chemical or microscopical examination of the urine, blood, fæces, and contents of the stomach about four hours after meals. The stomach should be free from fermentation organisms, and the stools show no undigested food or unusual fœtor. The blood should become constantly richer in red corpuscles, or in hæmoglobin, or in whatever element it is found defective in the first examination. Eosinophile cells diminish in number, poikilocytosis becomes less and less marked, and the flat red corpuscles grow fuller and more biconcave. The changes in the blood from day to day form a very good index of assimilation. While the patient is on an animal diet, the presence of indican in the urine (if there be no pus in the body) points to intestinal putrefaction, indol being a product of the putrefaction of albuminoids. The information obtained in this way is at once practical, scientific, accurate, and sure ; in a large clinical experience it has proved to be a satisfactory guide.

The application of these general considerations to the treatment of the special forms of the chronic disorders of digestion may now briefly command our attention.

In dyspepsia with increased formation of hydro-

chloric acid the patient must be held strictly to a
diet of lean meats. The keeping, selection, and
cooking of meats cannot be discussed in the limitations of this chapter. All lean meats should be
broiled or roasted, never stewed or fried. The staple food should be the muscle pulp of beef scraped
or chopped free of fibrous tissue, steak, roast beef,
or mutton chops or roast mutton. For the sake
of variety one can ring the changes on the white
meat of poultry plainly cooked, fresh white fish, or
raw oysters (care being taken not to swallow the
tough part) served on half-shell with lemon, or the
white of egg cooked just enough to hold together.
The juice of a few tender sprigs of celery or of
watercress or of horseradish, extracted with lemon
juice, may be used to give flavor. In the way of
drinks, a small cup of black coffee (if there is no
contra-indication) after breakfast and dinner, and
a small cup of clear tea at noon, should be recommended ; but no wines or alcoholic drinks what-
soever. As soon as healthy secretion is restored,
the crust of French roll, stale bread dry toasted,
and a few tablespoonfuls of well-cooked rice or
cracked wheat, or California wafers served with a
little butter and salt, and, a few weeks later, spinach, fresh English peas, string beans, a floury
potato, may be added as the patient is cautiously
conducted on the way to a normal diet. The juice
of ripe fruits may now also be taken without harm.

In dyspepsia with diminished formation of hydrochloric acid, and also dyspepsia with impaired
muscular movements, a diet of animal food should
be ordered until there is no longer any evidence

of fermentation, and the patient be then slowly brought around to a normal diet. The crust of roll, or stale bread toasted so dry that it will snap, are peptogenic, are more easily digested than starch. are not so liable to ferment, and may be given along with lean meat in the beginning of treatment. A few tablespoonfuls of bouillon before dinner will also increase the secretion of pepsinogen. Animal fat—as butter, or a slice of the boiled side of bacon, or cod oil—should be given as soon as the stomach and intestines are free of fermentation, to aid in toning and building up the central nervous system. But if the fat denudes the tongue or encrusts it with a layer of dead epithelium, or excites nausea or eructations, it must be at once withdrawn. Inunctions of animal fats or pancreatized cod oil as a nutrient enema may then aid. A glass of hot sterilized milk will often prove of value when sipped very slowly in the interval between meals, or at the beginning of the meal as a soup. The tea may be made more delicious by a slice of lemon and a teaspoonful or two of old velvety rum. A little old whiskey or brandy may be permitted if the heart is weak. The rule to return to a mixed diet suited to the state and needs of general nutrition, as rapidly as the capability of the stomach will permit, here also obtains. In dilatation, soups and milk do not agree ; the small bulk and high nutritive value and digestibility without irritation make lean meats the staple food. Fats must be watched.

In dyspepsia with exaggerated peristalsis the diet must be bland and unirritating. Milk and its preparations, lean meats, and light farinaceous food, without succulent vegetables and condiments, should be ordered until the condition is controlled by drugs.

In dyspepsia from sympathetic disturbance the diet should be fluid and non-irritating—as peptonized milk or milk gruel, koumiss, matzoon, buttermilk, white of egg, the juice of beef or other meats —while the disease of which the dyspepsia is a reflex is discovered and palliated or cured.

In chronic gastritis clinical experience has taught us in the beginning of the treatment to withhold starches, fats, and sweets ; and the less the chances given for fermentation and putrefaction the sooner we may expect a cure. The treatment proceeds along the same line as in weak stomachs, but progress is slower and minute attention must be given to every detail of management and every aid be brought to bear. In venous stasis, the stomach being kept clean, the diet should be such as will least irritate, and only enough albumin and fat to maintain the nutrition of the body be given. If the liver is involved, fat must be supplanted by carefully selected cereals and fresh vegetables. In threatened cardiac insufficiency, after diminishing the work of the heart and prolonging the period in which it may take its rest, give, along with the albuminoids and proteids, enough carbohydrates to enable some of the albumin to be organized, and thus guarding also against the storage of fat. The diet of dilatation has already been given, and artificial digestion is the only additional indication afforded by gastric atrophy. Many details have been written at the risk of becoming tiresome, and many more must be left to the good sense of the physician.

The diet of gastric ulcer must be unirritating to the lesion of the mucous membrane and adapted to the hydrochloric superacidity. It is essential to healing that the ulcerated surface should be given rest and that distention of the stomach be carefully avoided. An exclusive milk diet, since its use and recommendation by Cruveilhier, has given some very brilliant results. Milk, sweet and fresh and partially peptonized, and rendered alkaline by lime water or (better) by calcined magnesia or the lactate of magnesia, may be a good diet to begin with, administered in small quantities every three or four hours. It is unirritating and gives the stomach little chemical and motor work to do, since it is almost entirely digested in the duodenum. But milk cannot be given in sufficient quantity to maintain nutrition without stomachal distention ; and when, on account of the superacidity, it does not agree (as is often the case), it does a good deal of harm. It requires nutrient enemata to supplement it, is a very treacherous food, and no one can tell beforehand when its casein is going to coagulate in clots, sour, and decompose. Leube, in a very large experience, has obtained the best results from his sarco-peptones (prepared also by Rudisch, New York). This preparation is concentrated, nutritious, and unirritating, but very unpalatable. Preference might be given to Mosquera's beef meal and peptone jelly, on account of their greater pleasantness to taste and smell. Meat juices, white of egg, and fine muscle pulp of beef, with fresh pepsin, are very valuable and counteract the superacidity by combining with the free hydrochloric acid. We are seldom able to accomplish much by rectal feeding, but must resort to it when the stomach is much disturbed and during haematemesis. Rest in bed and the curative treatment of the superacidity meet two very important indications. The strict diet must be continued for some time after the disappearance of all symptoms.

The treatment of cancer is purely symptomatic. A restricted, mixed diet may be comfortably taken care of in the beginning, but as irritability, dilatation, and atrophy develop the diet becomes more and more exclusive, until finally sepsis and inanition or peritonitis close the scene.

Before passing from the dietetic to the strictly medicinal treatment something must be said on gastric cleanliness. We all know the radical revolution that cleanliness, antisepsis, and free drainage have created in surgery. The supreme indication in the treatment of an inflamed surface is to keep it clean ; and this is pre-eminently so when the inflamed surface is a highly organized secreting and absorbing structure. Stomach-washing is a crude attempt to apply these principles of surgery to the treatment of the diseases of the stomach. In dilatation of the stomach, or gastritis with the stomach irritated by the products of micro-organisms, in superacidity and supersecretion, it does not require much experience to teach one its value. It is best to employ the method on retiring, or in the early morning one hour before breakfast. The patients soon learn to do it themselves, or a nurse may be employed. Plain boiled water does the work well, to which, in hyperæsthesia of the mucosa, a little chloroform water may be added, and thirty grains of subnitrate of bismuth be given at the end of the procedure. Stomach-washing is palliative, but not directly curative, and is liable to be very much abused. It is often a better plan and more physiological to stimulate the stomach, when possible. to empty itself. by hot water, massage, and electricity. Change of diet, by changing the culture soil where fungi are present, also makes for cleanliness and is very refreshing. Drinking of hot water, by washing and draining downward, is also an important aid. It also soothes the terminal nerves and promotes secretion, just as the daily bath leaves the skin in a healthier state. Moreover. it is a gentle, safe, and sure diuretic, increasing not only the urine water, but also the solid excreta held in solution, promoting both the waste and renovation of tissue by quickening the circulation of the fluids. Hot water (it is needless to say that it should be taken slowly and at the right time) stimulates the heart and raises arterial tension without contracting the arterioles, and, consequently, should not be taken in hæmorrhage, menorrhagia. endarteritis, or in valvular heart disease when compensation is on the verge of being disturbed. Its diuretic power makes it the most valuable means we have of eliminating from the system the poisonous products formed by the tissues or absorbed from the alimentary canal.

The drugs most useful in these diseases are such as increase the capability of the stomach. promote healthy nutrition, relieve grave symptoms, or combat the morbid process. A sweeping exclusion may be made of all medicines that irritate the mucous membrane or derange the chemical process of digestion. Dosing with nauseating mixtures can only do harm. Drugs should be given with a definite purpose in view, and our aim in prescribing should be to combine simplicity, elegance, and power.

The capability of the stomach may be increased in many ways. In subacidity the most plausible thing to do is to give hydrochloric acid. It is a temporizing expedient, and, if it does any good whatever, certainly has no curative value. If administered it should be given in small doses repeated three or four times during the active period of peptonization, to avoid producing artificial superacidity. It is more rational and curative to excite secretion by massage, faradization, and drugs. The administration of alkalies in superacidity suppresses a symptom temporarily, but afterward excites acid secretion; to relieve the pain it is necessary to use, along with the alkali, an analgesic. Calcined magnesia and the lactate of magnesia are preferable to the alkaline carbonates, although "soda-mint tablets" are popular and also efficient. The curative treatment strikes at the cause by diminishing the irritability of the mucous membrane from which the supersecretion results. The fluid extract of coca (P., D. & Co.), the tincture of piscidia erythrina in small doses, and the English or Squibb's extract of cannabis indica, are the three reliable drugs for this purpose. Papoid is of value when given before the hot water to aid in the removal of mucus.

Then one or more of the following drugs, on account of their physiological action, may be selected to meet the varying indications of defective secretion and impaired movements: The simple bitters increase the acidity of the gastric juice, and are supposed to diminish the secretion of mucus; the proper time to administer them is half an hour before meals; all of them are local irritants, and their use should not be continued longer than three or four weeks. Ipecac promotes the secretion of mucus, and in small doses allays irritability. Opium, morphine, and codeine diminish acidity, allay irritability, and check peristalsis without affecting absorption. Nux vomica increases the acidity of the gastric juice and tones and strengthens the muscular layer. It is the one drug to use in dyspepsia with diminished muscular movement. It also increases the quantity of nerve force radiating throughout the body, and this important action may often be used to promote tissue building. If too long continued the discharge is excessive and waste of tissue results. Subnitrate of bismuth is astringent, antiseptic, and sedative. Nitrate of silver allays irritability and is supposed to exert a specific action in catarrhal inflammation. Arsenic inhibits the activity of the hepatic cells, and is prescribed empirically in the neuroses; in the neuroses of the stomach Fowler's solution in drop doses, before meals, is of some value; or the bromide of arsenic or of potassium or of sodium may meet an indication. Iron, the alkalies, oxalate of cerium, and the stimulating antispasmodics are at times of value: also calomel, cascara sagrada, ipecac, aloes, rhubarb,

senna, and podophyllin are useful to gently touch
the liver or to keep the bowels in a proper state.
Salicin, chloroform, and camphor are antifermenta-
tive, but the best way to prevent fermentation is to
keep the stomach clean, give the proper food, and
see that enough hydrochloric acid is present.

To discuss every indication in the treatment of
these diseases of the stomach would be to write a
volume on therapeutics. To summarize, in conclusion :

1. Chronic gastritis is rarely, and dyspepsia almost never, a primary local disease. Ulceration is
a local trouble engrafted on a secretory neurosis and
a blood condition. Atrophy may result from local
or constitutional disease. Cancer may be primary,
or, rarely, is secondary.

2. An accurate diagnosis means more than the
discovery of defective gastric digestion. We must
know the anatomical state of the mucous membrane.
We must also know the nature of the disturbance
—whether of secretion, movement, or both ; the
source of the disturbance—whether in bad habits of
life, in acquired or inherited defect of constitution,
in vice of nutrition, in fault of elimination, or in
disease of a distant or functionally associated organ.
The solidarity of the organs of digestion is a fact
of very great importance in clinical medicine, and
dominates the method of managing their disorders
and diseases. Their intimate relation through a
common nerve supply ; the mingling in the portal
vein, on its way to the liver, of the various materials
absorbed from the alimentary canal ; the division
and community of their labors ; the integration of

their differentiated functions, make them one in action and in purpose.

3. The treatment embraces more than the management of the local disturbance. The local treatment is important; the stomach must be kept clean and sweet, its work diminished, its capability increased. But the whole man commands pre-eminent consideration—his mental, moral, and physical condition. And this necessitates the study of the character of the patient, the regulation of his habits of life, the prescription of palliative and curative remedies, and a well-regulated diet. And a well-regulated diet does not mean the arbitrary and indiscriminate use of certain articles of food, but a diet sanctioned by reason and experience, adapted to the state and needs of general nutrition, and to the capability of the stomach and to the peculiarities of the patient. But of more importance than all else is the complete digestion of the food taken : this the physician must see to by daily observation, little changes in quantity, quality, or frequency, and by wearisome and prolonged supervision. The mere suppression of symptoms will do the patient no permanent good. It is better to restore than to supplement secretion, and to correct than to neutralize superacidity. The curative treatment is directed against the chemical lesion of the fluids of the body and the malnutritive state of the cellular protoplasm.

CHAPTER III.

In the clinical study of the disorders of digestion the stomach cannot be considered the most important division of the alimentary canal. In the light of modern research this position must be assigned to the small intestine, and chiefly to its upper part. It is in the duodenum, and in the duodenum only, that a mixed diet can be perfectly prepared for absorption. The work begun in the kitchen and continued in the mouth and stomach reaches the climax of chemical changes at this point. The preceding stages of digestion have been preparatory and progressive.

Duty and responsibility go hand-in-hand. When the duodenum with its two great appended glands was supposed to play a subordinate and supplementary part, not much attention was given to the intestine in the disorders of digestion, and the logical sequence was failure in treatment. The stomach has been much abused by laymen, and a physician of genius has seen in it the origin and source of every form of chronic disease. No other organ has been so maligned and maltreated. It is now time that the responsibility should rest where it belongs, and much of the blame must be transferred to the intestine. Vicarious suffering is not a principle of law or of Nature or of disease.

The stomach is an antiseptic receptacle which doles out its contents to the duodenum in a soft, semi-fluid, mixed, and slightly changed form. Its secretion, as does the saliva, only acts on one class of foods and in a very incomplete manner. No very great quantity of nitrogenized food is converted by hydration into peptones, and the precipitated casein, proteoses, liberated granulose, and fat are discharged into the duodenum. But be it understood that it is not our purpose to underestimate the utility of the work done by the stomach. There is much reason for believing that it would be disastrous to have all of the proteids and albuminoids converted into trypsin peptone, which is essentially a decomposition or erosion product, and one form of which is utilizable only in the production of energy and animal heat. Gastric peptones can be readily converted by anhydration into serum albumin and are available for tissue building. Pure peptones suffice to keep up nutrition (Maly, Adamkiewicz). Moreover, unchanged albumin introduced into the rectum is absorbed and can maintain nutrition (Ewald and Eichhorst), and proteoses are even more readily drawn into the circulation. Incomplete peptonization cannot, therefore, be admitted as an argument against the usefulness of the work of the stomach in digestion. Careful alimentation can maintain nutrition in the dog and in man without the intervention of a stomach. This proves that the work of the stomach is not essential and can be delegated, in certain favorable conditions, to the duodenum. It detracts not one iota from its value, and the richness of resource results from the development in duplicate and the multiple relation of function to structure in the evolution of the digestive system. The stomach also does important police duty in destroying pathogenic bacteria and ejecting indigestible, irritating, and poisonous substances. The cardia and pylorus open and close opposedly. The eyelid, by a beautiful provision of Nature, protects the organ of sight. The muscular pylorus holds the door to the intestine. But the chemical work of the stomach is not all-important, and this pouch is simply an antiseptic, protecting, distributing, and chiefly preparatory receptacle.

The intestine is a digesting, absorbing, and eliminating tube. Our study is restricted to the disorders of digestion, and absorption and elimination can only receive consideration as causative factors. Elimination may disorder the digestive process by altering the chemical reaction of the intestinal contents or by originating a diarrhœa. If the intestinal epithelium loses its selective power, auto-infection, with its pernicious influence on the system and on digestion, may result. An excess of the diffusible products of digestion interferes with the further action of the ferments, and deficient absorption predisposes to superdigestion and organic putrefaction and fermentation.

The part that the secretion of Brunner's glands plays in the conversion of the food into a liquid and diffusible product is not well known. This juice liquefies proteids and albuminoids, acts vigorously on a ptyalin product, maltose, and probably also on cane sugar, and by its intense alkalinity aids in the

neutralization of the gastric juice. Its defective
secretion may add to the work that must be done
lower down in the alimentary canal, and we would
naturally ascribe to its absence a predisposition to
the simple duodenal ulcer.

Incomplete also is our knowledge of the enteric
juice. Cane sugar is chiefly prepared by its inver-
tin, and its alkaline carbonate is essential to the
maintenance of the proper reaction of the contents
of the jejunum and ileum. Without it the pancreatic ferments would soon be rendered inactive by
organic fermentation and putrefaction, although
normally no bacterial decomposition of proteids and
albuminoids should take place in this part of the alimentary canal, and indol and phenol and scatol and
marsh gas and hydrosulphuric acid should never be
formed here in health. The gastric hydrochloric
acid should be neutralized in the duodenum, and
consequently the intestinal contents are nearly neutral in the jejunum but always faintly acid in the
ileum. It is the organic acids that the enteric
juice is called upon to neutralize, and the maintenance of a proper chemical reaction has a good deal
to do with the prevention and limitation of bacterial growth.

The diastatic ferment of the bile acts feebly on
carbohydrates, and the bile salts throw down the
proteoses in the form of a flocculent precipitate.
The digestion of a natural emulsion of fat (cream)
is perfect in the absence of the pancreatic juice
(Dastre), and about ninety per cent of it is absorbed
(Mering, Minkowski), but not so in the absence of
bile. Without bile neutral fats are not emulsified,

4

and organic decomposition is less when bile is present. Its antiseptic properties are very feeble (though it seems not to be a very good food for bacteria), and it exerts its favorable influence by promoting pancreatic digestion, absorption, and peristalsis. The liver is of greatest use in metabolism.

The pancreatic juice puts the crown on the chemical process of digestion, and its work gradually loses itself in organic decomposition. It prepares no way, is regal in its advancement, but its rule is limited by precedent and hedged about with chemical and vital law. It is with this code that we are chiefly concerned.

Perfect duodenal digestion requires (*a*) a medium of proper reaction, (*b*) normal secretion, (*c*) a proportionate quantity of digestible food in a proper physical condition, and (*d*) the normal movements of the food mass.

It may be supposed that the best reaction for the food mass to possess is the one which is most favorable to the action of the digestive ferments—the trypsin, amylopsin, steapsin, and milk-curdling ferment. In perfect health this is probably true; in disordered assimilation rapid digestion and rapid absorption may both be undesirable. But our study is limited to disordered digestion, and it is our purpose to consider the changes in the environment and in the conditions which disturb and delay the process. It is well known that the pancreatic ferments are most active in a slightly alkaline medium. The essential condition is complete neutralization of the hydrochloric acid. In the presence of bile a feeble acidity due to organic acids does not inhibit but probably increases their activity (Lindenberg). The chemical equilibrium may be destroyed by a too acid chyme, by a deficiency of the duodenal secretions, by excessive organic fermentation, and by too little enteric juice. The excess of acid may be taken in the food, or it may be developed by organic fermentation or fat-splitting in the stomach, or it may be the result of excessive secretion of hydrochloric acid when the pancreatic ferments are not only rendered inactive but are also destroyed.

Duodenal dyspepsia from defective secretion is a frequent disorder. There may be too little pancreatic juice or too little bile, or there may be too much bile of a bad quality, producing excessive peristalsis. Normal chyme is probably the best stimulant of duodenal secretion. There is the same orderly sequence in secretion as in the digestive process. Through nervous association salivary is followed by gastric secretion, and then the duodenum and its appended glands are aroused to action. The alkaline saliva promotes the secretion of the acid gastric juice, which in its turn puts the duodenum to work.

Duodenal acidity and faulty secretion are not the only disturbing factors, but the chemical process in the intestine may be disordered by an improper composition, or faulty preparation, or excess of the chyme. Gluttony is a frequent cause. An excess of proteids or of carbohydrates or of fats is no less pernicious in its ultimate effects. Either form of excess throws too much work on the duodenum, which will inevitably become inadequate. Not only is the influence direct, but indirect also through defective preparation by the mouth and stomach. The result, however brought about, is a chyme abnormal in quantity or quality.

The intestinal wall contains two sets of muscular fibres which are often dissociated in their action—the one regulates the calibre of the gut, the other the movements of its contents. Peristalsis and tonicity often act in unison, but just as often apart from each other. Hot water increases peristalsis (Ricord), but diminishes tonicity ; cold water increases tonicity and may or may not influence peristalsis. The dilated stomach spasmodically empties itself, and the same is also true of the dilated colon. The investigations of Glénard show very plainly that hypertonicity and inadequate peristalsis coexist in enteroptosis. The habitually relaxed pylorus often allows the food to be hurried into the duodenum. Neurasthenics often have flat bellies, cord-like intestines, and constipation. And it is important clinically to remember that these two kinds of muscular action may be variously combined and localized, and restricted to divers parts of the digestive tube. Hypertonicity disorders digestion by diminishing the area of absorption and interfering with the circulation of the blood. The food mass is not churned and brought into ever-varying contact with the mucosa. Insufficient and irregular and excessive peristalsis delays and disorders and decreases digestion and absorption. Atonicity permits stasis. Perfect digestion requires normal chemical and muscular action. The physical factor is no less essential than the chemical one.

The recent brilliant discoveries in the chemical process have drawn our eyes away from the muscular layer. Unhealthy variations in intestinal tonicity and peristalsis are probably more pernicious in their influence than defective duodenal secretion.

From these proximate causes turn we now to the consideration of the remote ones. Digestion is disordered by every disease which is not purely local in its nature and effects. And our knowledge would naturally lead us to expect this, since perfect digestion requires, in addition to a right quantity of healthy food, normal nerve centres, a normal supply of pure blood, normal secretory and absorbing cells, and normal tonicity and peristalsis. These conditions are incompatible with every disease which is not strictly local and which involves a part that is not a component of the digestive system, be that disease discoverable with the microscope in the destruction, arrangement, or production of cells, or hidden under the word "functional" in intracellular change. The neuroses, denutrition, alcoholism; anæmia, chlorosis, malaria, and other forms of toxæmia; organic disease of the hæmatopoietic or metabolic, respiratory, eliminatory, circulatory, or nervous systems, or of the digestive tube and its appended glands, may be the ætiological factors. To enumerate the remote causes of intestinal dyspepsia would be to pass in review the entire number of chronic disorders and diseases capable of disturbing one or more of the conditions of perfect digestion. If we carefully consider the clinical history, the subjective symptoms and their order of development, the physical

signs, and the result of the chemical and microscopical examination of the blood, secretions, and
excretions, we will commonly be able to adopt a
rational supplementary treatment directed against
the remote cause.

The symptoms of intestinal dyspepsia are constitutional and local; the two symptom groups are
born and develop and live and decline and fall together. We are well aware that we are now tread-
ing on disputed territory; the battle yet rages
fiercely and the existence of neurasthenia and this
great class of dyspepsias is staked on the issue.
Specialism has joined the fray, and the war is to
the knife. Are these symptoms, including those
that are localized in the digestive tube, due to neurasthenia, to a functional nervous state without
anatomical change (Beard), or to hyponutrition of
the nervous system (Arndt), or to a general neuro-
pathy affecting alike the digestive tube with all organs (Charcot), or to dilatation of the stomach with
auto-intoxication sequential to chronic gastritis
(Leube), or to weakness and relaxation of the muscular layer (Bouchard), or to dilatation of the as-
cending (Bouveret) or descending (Trastour) colon,
or to enteroptosis (Glénard)? These questions can
best be answered at the bedside by the general
practitioner. His is the eagle eye that sweeps the
whole field in a flash and takes in every detail. The
vision of specialism is all the more intense because
of its exclusiveness, but on broad questions is very
apt to be wrong because perfect truth comes full
circle. It seems probable that the neurologist and
specialist in the disease of the digestive system,

though diametrically opposed, are walking in the
same beaten pathway, in the same vicious circle,
which was long ago established when nutrition,
circulation, and the nervous system were linked
together in the one law of being. It may be the
nervous system that is robbed of its food and rest,
and brought to a premature fall by hard hunger and
an overreaching ambition; it matters not whether
the force be scattered in the shock of the lightning
flash or slowly wasted beneath some burning ray,
the result is the same—a nervous wreck more or
less complete. The beginning may have been small
—a slight malaise. The end is complete prostration. And associated with the gradual decline or
the rapid fall are divers disorders of the digestive
process. Neurasthenia is one of the grand causes
of gastric and intestinal dyspepsia, and affects primarily and chiefly the neuromuscular factor, the
physical process. Associated with it there may be
normal secretion (or even hyperchlorhydria) or deficient secretion. There may be hypertonicity with
a small stomach and cord-like intestine, or there
may be flaccid dilatation. But there are essentially
and primarily diminished peristalsis and constipation, and sometimes complete stasis. Now, it is the
digestive system that first fails, and the primary
disorder is in the chemical process, as is usually the
case also when there is "somewhat wrong with the
blood." Neurasthenia is an entity; so is intestinal
indigestion. The one may cause the other. Each
may exist alone. Both may result from a common
cause. Both are parts of the same circle, which
often becomes a vicious one. What, then, are the

symptoms of intestinal dyspepsia, and on what can its diagnosis be based with certainty ?

Habitual malaise and general debility are the two earliest and most persistent symptoms. A little work easily tires ; sleep does not refresh ; the mind is uncontrollable, wandering, flighty. The thinker cannot concentrate his attention ; thought loses both in intensity and extension. The broad view and firm grasp require a supreme effort which leaves relaxation and exhaustion. The philosopher becomes gloomy and apathetic, or pessimistic and crabbed. The preacher grows ascetic and the brightness of hope is replaced by the gloom of despondency. The poet loses some of the sweetness and clearness and continuity of his song. The artist fails in conception and trembles in execution. The musician turns from his instrument—cannot rest, cannot compose. The statesman becomes sour and oppressive and defiant. The merchant is swallowed up in competition. Poet and plowman, priest and philosopher, one and all, lose energy, pertinacity, strength, and happiness because the intestine does not do its work well, and the liver gets clogged, and the blood contaminated, and the nerves irritable and tired and without reserved store of force. Probably neurasthenic first, dyspeptic afterward—the vicious circle is established, and neither rest nor diet alone, but both combined, will cure. The malaise is worse a few hours after meals ; the general debility is most felt after a little forced work ; both are usually at their height about the middle of the afternoon. Habitual malaise and general debility begin and rise and decline and fall with the disorder of digestion.

Insomnia, in many cases, is a most obstinate symptom, and most frequent in the early morning hours. Alcoholic drinks aggravate it, and the only hypnotic that will give refreshing sleep is a clean digestive tube.

Sensory disturbances are frequent. Neuralgia, hyperæsthesia, paræsthesia, anæsthesia, even lancinating pains like those of locomotor ataxia, are not rare. These symptoms bear no definite marks, and are mentioned only on account of their association with, and proportionate relation to, the degree of the digestive disorder.

The heart symptoms are reflex or mechanical or due to auto-infection. Tachycardia, which may be paroxysmal, is not rare. The heart muscle is nearly always weak and the peripheral circulation poor. Vertigo from cerebral anæmia or auto-intoxication is only too common. Palpitation seems to be about as often found as in gastric dyspepsia. But the chief cardiac sign is the condition or behavior of the right ventricle. Flatulence, especially in the transverse colon, interferes with the action or filling of this ventricle, and the heart is pushed up and laboring or rapid, the respirations are quick and shallow, the pulse small and compressible, and the veins are full. The dyspnœa may be increased by the clogging of the liver, auto-infection, and contraction of the pulmonary arterioles. The symptoms may be intermittent or remittent or paroxysmal, accordingly as may be the strength and adequacy of the right ventricular wall. The heart may be not only inadequate but also irregular. The diagnosis of dilatation of the right ventricle is not difficult, and the therapeutic test of the relation of the cardiac trouble to the disorder of digestion is conclusive. Treatment directed to the heart alone fails. Digitalis and drugs of a similar nature do harm. Strychnine and nitroglycerin aid, but alone are inefficient or useless; but, combined with rest and a diet to control flatulence and to cure the intestinal dyspepsia, will sometimes restore the equilibrium, even when the heart is near the stage of asystole.

Distress and pain and tenderness are among the *local* symptoms, but cannot be considered as pathognomonic. The central figure on the canvas does not make the complete picture, and it gets a good deal of its meaning from its relations and associations: two peasants standing with heads bowed in devotion may not attract more than a passing recognition, but the dropped work, characteristic scenery, and sound of the distant church bells wake into expression a grand and touching historical truth. It is not on any one sign, but on the symptom group, that our diagnosis must rest. Very little meaning can be attached to the time of appearance of these symptoms. Their location should be considered. But the most valuable sign is a bruised and heavy feeling in the belly during the restless hours of the early morning.

Persistent flatulence in the small intestine is an almost pathognomonic sign of intestinal dyspepsia. It is greatest when organic putrefaction and fermentation are most active, and this usually occurs two or three hours after a meal. It is by no means rare to have gas diffused from the blood into the intestine, but this occurs irregularly and intermittently, and chiefly when the intestine is empty, and is not related to the quality of the diet. When poured into the duodenum from the stomach the clinical history and physical signs will suggest its source, and the urine and stools will contain nothing indicative of intestinal indigestion and decomposition.

Dilatation and displacement of the intestine is a physical condition and sign of some value. It may be due to either distention or relaxation; uneven tonicity, especially when combined with localized atonicity, may produce stasis of the intestinal contents; deficient peristalsis and chemical and bacterial decomposition mechanically distend. The flexures of the colon finally are displaced and fall from lax ligaments and a flaccid abdominal wall. This condition develops *par excellence* in the neuromuscular form of dyspepsia.

Constipation and irregular stools vary with the quantity of the bile, the chemical and physical qualities of the intestinal contents, and the disorder of the muscular layer. Organic acids, scatol, carbonic acid, hydrosulphuric acid, and marsh gas excite peristalsis; nitrogen, hydrogen, indol, and phenol have no influence (Bokai).

The urine is more or less characteristic. Indol is formed by the decomposition of tyrosin, a product of trypsin superdigestion, and by the bacterial decomposition of nitrogenous compounds, and it appears in the urine as indican. This process normally never occurs in the small intestine ; and a urine containing an excess of urates, occasionally a few crystals of uric acid, of specific gravity about 1.020, a trace of bile, and indican in excess, is almost pathognomonic of intestinal indigestion, if the large bowel has been previously washed out. The deficiency of acid in the urine gives some idea of the amount of HCl secreted (Ewald), provided the increased alkalinity of the urine is not due to the absorption of alkalies from the food (Roberts), or to loss of HCl by vomiting, or to delayed absorption after secretion, or to the formation of insoluble chlorides (Jones and Quincke). This is a more trustworthy index if the neutral or feebly acid urine precipitates the earthy phosphates on boiling. The alkaline secretions diminish the alkalinity of the blood and increase the acidity of the urine (Hübner, Sticker, Jones, and Quincke). An excessively acid urine of normal or high specific gravity, and which, after standing forty-eight hours, only deposits, it may be, a few crystals of uric acid or oxalate of lime, is produced in this way. In hyperchlorhydria the abstraction of acid is followed by the withdrawal of alkali in excess to neutralize it, and the reaction of the urine is unchanged or vacillates. Excessive organic fermentation and consequent excessive secretion of the alkaline intestinal juice are the conditions underlying the formation of the clear, highly colored, excessively acid urine which very much delays deposition.

The stools are often characteristic from the fermentation and putrefaction to which they testify,

or from the excess of unutilized starch and fat
which they contain.

The diet test is the sure proof, and is based on the
intolerance of starches, fats, sweets, and wines.
Milk consequently is one of the first of the common
foods to disagree. Starches, unless permitted to be
destroyed by stasis and fermentation, are voided in
excessive quantity. Fats escape in like manner in
the fæces. Sweets add proportionately to the flatulence. All wines, except the oldest and lightest,
are badly tolerated. Make carefully selected and
scientifically prepared and easily digested and nutritious meats the basis of the diet, give one or more
of the badly tolerated class of foods in an easily digested form and not in excess, regulate peristalsis,
examine the stools, apply our knowledge of physiological chemistry, and the results will be pretty definite and conclusive.

Such are the particular symptoms of which the
symptom group is composed, and it is on the ever-
varying combination that the diagnosis of intestinal
indigestion is based—a diagnosis which is always
difficult and requires the very closest clinical study.
The chemical condition of the stomach, both during
and in the interval of digestion, the time and thoroughness with which it empties itself, its size and
the tonicity or flaccidity of its walls, can by a few
examinations and tests be readily ascertained with
a good deal of certainty. But the disorder in the
intestine is enshrouded in difficulty and well protected against chemical exploration. But a meth-
odical study of the symptoms and of the physical
signs, the examination of the urine and of the

stools, and a careful use of the diet test, will make it possible to form a right and definite conclusion. To each symptom we assign its possible causes — what conditions and where located would produce it. In turn we treat each prominent symptom in this manner. We then apply the same method to the symptoms as combined until we arrive at the possible explanations of the symptom group. In this procedure the chemical or physical process of digestion will be found more or less faulty, and possibly also the special defect be revealed. The examination of the urine for decomposition products, after the large bowel has been previously thoroughly washed out, will confirm or further limit our conclusions and supplement our knowledge. The diet test may then be made, and a positive result will give to our inferences a high degree of moral certainty. This method will turn on more light than any other with which I am acquainted, but it requires time, close observation, careful reasoning, and disagreeable work. The solution of a difficult problem and the rational treatment of the patient are the rewards of the conscientious endeavor.

It remains to differentiate intestinal from gastric dyspepsia, and then to separate the disorder into its three great varieties. But be it understood that certain forms of gastric dyspepsia always lead to disorder of the duodenal process, and, *vice versa*, that intestinal indigestion frequently deranges the functions of the stomach ; and that the two are sometimes inseparably bound together as the manifestation of a common cause or as the expression of one disease.

Heartburn, acidity, pyrosis, nausea, vomiting, epigastric pain and tenderness, are more or less characteristic of gastric dyspepsia. Flatulence can be located in the stomach and in the intestine by the physical signs. The time of appearance of the distress or pain must not be given too much consideration and value ; the pylorus is not an incorruptible guard ; gastric peristalsis is not a fixed quantity. The food does not, like a sparrow—to adopt a favorite simile of early English song—fly in at one window and, after a brief sojourn, disappear through the other. The entrance is usually rapid and surprisingly abrupt, at least such is the custom in America ; the duration of the rest is very variable, and the time of departure of each individual traveller is conditioned by varying circumstances. Nothing is more remarkable than the likes and dislikes, the whims and fancies and conduct, of the human stomach. If it be remembered that the stomach can be filled with swallowed air or with gas regurgitated from the duodenum or diffused from the blood, the time of appearance and location of the flatulence, pain, and discomfort will be available in differential diagnosis. Auto-infection is more common in intestinal indigestion. It may well be doubted that even in the flaccid gastric dilatation of Bouchard the toxines are formed in the stomach and enter the system from this point, as the neuromuscular form of intestinal indigestion is the usual accompaniment of this condition. Simple emaciation without cachexia, or a full and ruddy face with vaso-motor unrest, is the rule when the disorder is limited to the stomach ; the muddy com-

plexion of severe cases of intestinal indigestion is
well known. The urine is sometimes characteristic ; the diet test is of inestimable value ; and the
physical signs of gastric dilatation, and of dilatation
or contraction of the colon, may be of very great
weight. It is not so easy a matter as might be supposed to diagnosticate and locate dilatation. In
using inspection, palpation, and percussion it is essential to remember the surface anatomical mark-
ings. About five-sixths of the stomach lies to the
left of the median line in the epigastric and hypo-
chondriac regions, and is entered by the œsophagus
behind the sternal insertion of the cartilage of the
seventh rib ; the pyloric extremity (about one-
sixth) is to the right of the median line, and terminates in the duodenum on a level with the tip of
the ensiform cartilage, and about two inches to its
right, behind the end of the eighth costal cartilage.
When gently distended the fundus rises to the level
of the fifth rib, and the greater curvature sweeps
forward and downward to the right, passing just
above the umbilicus. It is easy to see how the
overdistended stomach produces dyspnœa and palpitation by interfering with the action of the right
heart and diaphragm and the expansion of the
lung. The cardiac end is fixed, the lesser curvature is only slightly movable, and the position of
the greater curvature is conditioned by the degree
of distention of the stomach and the displacement
of the pylorus, which in disease can sometimes be
felt below the lower border of the liver. Only a
small area of the organ is superficial and in contact
with the abdominal wall below and beyond the left

lobe of the liver and with the left anterior thoracic
wall, the latter forming the half-moon-shaped space
of Traube. The colon begins with the blind pouch
in the right iliac fossa, ascends in front of the right
kidney and forms the hepatic flexure near but to
the right of the gall bladder, arches backward
across the abdomen above the navel in a line join-
ing the tips of the eleventh ribs, bends beneath the
lower border of the spleen, and descends to the
upper part of the left iliac fossa, where it terminates
in the sigmoid flexure. The large bowel is very
movable, the transverse arch is particularly free,
and the cæcum, the hepatic, splenic, and sigmoid
flexures are the favorite sites of dilatation. In the
diagnosis of gastric dilatation the methods of Fre-
rich (distention by CO_2 generated in the stomach),
of Lente (palpation by the sound moved about in
the stomach), and of others (pumping in air, to distend the viscus, through the stomach tube) are not
available in private practice. The clinical history,
the discovery of the peculiarly shaped asymmetri-
cal bulging on the left side and the perception of
peristalsis, the examination of the vomit, succus-
sion splashing and seething, the location by palpation and percussion of the greater curvature on a
level with or below the navel, will commonly establish the existence of extreme and moderate dilatation without a resort to heroic procedures. If, after
emesis or stomach-washing, a glass, or even a pint,
of water is introduced into the stomach, the line of
water dulness in the erect position, which is supplanted by resonance when the patient lies down,
will locate the lower limit of the stomach (modified

5

after Penzoldt). The pitch of the percussion note is higher in clonic dilatation, is commonly associated with large and foul diarrhœal movements alternating with constipation ; the dilated part can be flushed out with a saline purge and enema, and inflated with air through a long rectal tube ; and if the stomach is not dilated the vomit and clinical symptoms peculiar to gastrectasia are absent. It is on these considerations that the differential diagnosis is founded.

A classification for use at the bedside should be simple and each division clearly characterized by distinct symptom groups. The disorders of digestion may or may not have a basis in pathological anatomy. and morbid tissue change may underlie or accompany the unhealthy variations in the physiological process. We will, therefore, consider discoverable lesions as links in the ætiological chain, and classify intestinal indigestion accordingly as the *chemical* or *motor* process or *both* are disordered. The third is a union of the first two varieties, which are joined by a common bond, the one being dietetic or neurosecretory, and the other neuromuscular. There are two sets of nerve fibres (or one set having a double function) controlling secretion. the one influencing the functionating cells and the other the blood supply. The blood and the nerves, through their intimate relations with nutrition, commonly fall together, and it is chiefly a matter of historical or scientific curiosity as to which was first in the field ; when the patient consults the physician the two forces are usually closely allied in a self-destroying war.

A great deal has already been said under ætiology and symptomatology that is useful in the differentiation of the varieties, and the reader will be spared a repetition. We would add a few words on the " *diet test* " before passing on to the treatment. It is much more satisfactory and more definite and more conclusive to make a test tube of the alimentary canal than to try to imitate natural digestion in the laboratory. If the cause of the disorder is dietetic and the abuse or error has not established a motor or secretory defect, the restriction of the quantity and the regulation of the quality of the food which composes the mixed diet will relieve the symptoms and make the patient comfortable. If the chemical process is at fault the starches, fats, sweets, and wines are badly tolerated and imperfectly digested. The presence of dilatation, constipation, or diarrhœa would incriminate the motor factor, and, after its regulation, the toleration of the foods normally digested in the intestine would exclude defective secretion. And we read in the urine, in the fæces, in the physical signs and subjective symptoms, the result of the experiment which Nature, the master physiological chemist, has performed under our direction.

Digestion is accomplished in contact with, but virtually on the outside of, the body, and, as we have seen, can be deranged in two ways—by unhealthy alimentation and by faulty secretion and motility. A proper diet alone will effect a cure if the disordered chemical process has not established abnormal secretion or muscular movement. But cases so simple rarely come to the physician's office.

It matters little through what channel the digestion has been disturbed. If the cause is present and still active it is essential to direct our treatment also against it; but the damage persists after the removal of the cause. The origin of the trouble may be in improper eating, or in unphysiological living, or (if the gynaecologist will have it so) in a diseased uterus, tube, or ovary; but you may regulate the diet, put the manner of living on a right basis, and restore the generative organs to health or cut them out, and the intestinal indigestion will still persist as a most damnable and rebellious legacy. And so it is that when the nervous system and nutrition are brought under the evil influence, the only hope of cure lies in a comprehensive treatment that reaches out beyond the local causative and digestive disorders and embraces the patient, that secures good digestion, healthy nutrition, and physiological living in a suitable environment.

We have already seen how large a number of intestinal dyspeptics, through forced work or through hyponutrition, are or become neuropathic. And it is the neuropath who requires faith and hope and contentment to lead him on. Mind is a very subtle power which modifies in some unknown way the medium through which it arises and the parts to which it expresses its commands. Thought, feeling, and emotion are not simply the aurora of mysterious cerebration—the correlatives of material impressions. Man is not a mere automaton, conscious or unconscious, as heredity, development, and experience dictate. But the brain, in a sense, creates and controls the life of which it is the engrafted flower. The influence of the mind on function, particularly on digestion and nutrition, is very great. This is the thread of gold, the bright line of truth, which runs through many a grand error or delusion. Suggestion (or expectant attention), all unconscious though it be, is the wonder-working power in amulets, relics, magnets, in "Christian science," in the "faith cure," in hypnotism. Disbelief prevents or breaks the spell. *The full confidence and hearty co-operation of the patient the physician must possess in order to be master of the situation; and a hopeful, cheerful, contented mind is a power which makes for health.*

It is the business of the physician to instruct as well as to bless. To do the best that others have done and that he himself can think of for the relief or cure of disease is not the fulfilment of his high calling. The physician's office is a university hall as well. And the remarkable ignorance which prevails, among even the most enlightened people, of the plainest and simplest rules of healthy living, reveals only too clearly the manner in which these public duties are performed. Dyspeptics are as ignorant and perverse as little children, and we must first tell them how to keep well before directing them how to get so. A very large percentage of the disorders of digestion are either caused or nurtured by bad habits, and it is most useful and essential to enforce physiological living as regards bathing, eating, rest, exercise, work, sleep, clothing, mental and moral control.

A good morale, physiological living, and a proper diet comprise the treatment of the mild cases.

Benefit will also be derived from mild local and general faradism, massage and Swedish movements, outdoor life in a pure atmosphere, and general tonics. These patients with slight disorder of the digestive process are usually too much drugged. This overzeal on the part of the physician is to be attributed to the impatience of the dyspeptic. Permanent results come slowly. The digestive organs have been habituated to the performance of bad work, and it requires time to eat away the iron chains. It takes anywhere from three months to as many years to correct the unhealthy variation, which has an inherent power of self-perpetuation, and to make, through force of habit, normal digestion the law of being. Physiology and pathology diverge on a plane inclined downward, and progress becomes faster and easier every day along the route selected by circumstance. Law is supreme and irrepressible both in disease and in health, and we direct and fix the vital force in the right channel by the proper changes in the physical, chemical, nutritive, mental, and moral circumstances by which its action is conditioned. Not the relief simply, but the cure, of these chronic disorders of digestion requires time.

But in the severe cases the treatment must comprehend other remedies and meet other definite indications. The one general condition which rises above all others in its evil influence is self-infection. Careful alimentation and strong natural barriers (active oxidation and a good liver) will arrest or destroy, while active elimination will remove, the impurities and poisons. The most powerful eliminating agent at our command is water (pure, either at spring water temperature or hot) in large quantities. Self-poisoning is most frequent in indigestion accompanied by dilatation and deficient peristalsis—in the motor variety of the disorder ; in a mild form it is not rare in chronic chemical dyspepsia. It is well known how frequent an accompaniment it is of acute dyspeptic attacks, both when primary and when engrafted on the chronic trouble.

The special treatment of the disorders of the motor process includes many remedies of very great power—electricity, massage, stomach and colon washing, abdominal support, and drugs which give tone and strength and regular action to the muscular layer.

Faradism is the form of electricity that is of greatest utility. Central galvanization, when both secretion and motility are faulty, seems to pay for the time expended in its application. The anode is placed over the cilio-spinal centre and the cathode is pressed in over the solar plexus, and an uninterrupted current of about ten milampères passed during a short séance. Mild general and local faradization imparts strength and tone to muscles and nerves. Local faradization also excites and regulates secretion. One broad electrode is placed behind over the cardia or lumbar region and the other slowly moved all over the stomach, intestine, and liver. With the intragastric use of electricity I have no experience.

Massage, like electricity, strengthens the abdominal muscles, increases gastric and intestinal tonicity and peristalsis, improves the local blood and lymph circulation, and promotes secretion. The time, duration, and frequency of the sittings and rubbings are determined by their objects and the effect produced, each individual case and condition being a law unto itself. Both remedies are contra-indicated by inflammation, malignant disease, ulceration, and generally also by the active period of digestion.

Stomach-washing is a very popular remedial procedure. I find myself using it less and less every day. It is the remedy *par excellence* when there is spasmodic or organic stricture or obstruction of the pylorus. But in atonic dilatation the pylorus is yielding or already wide open. The stomach is then best cleaned and emptied by copious draughts of hot water, massage, and local faradization. This method stimulates and aids and encourages the organ to empty itself in the normal way. Stomach-washing, on the contrary, leaves the viscus clean but flaccid.

The same objection applies, though in a less degree, to washing out the dilated colon. Mechanical distention does not improve tonicity and peristalsis. The procedure is useful to secure cleanliness while we stimulate and encourage by massage, electricity, and drugs the weak and lazy bowel to the performance of its work.

Sulphate of strychnine, in minute doses, is beyond question the best drug for this purpose. Tinctures and wines and syrupy mixtures are objectionable. Coca and damiana may also aid. Aloin, ipecac, senna, rhubarb, or stronger purgatives may be required for constipation.

The abdominal or pelvic supporting band as a remedy in dilatation and displacement we owe to the genius of Glénard. It should extend high enough to support the stomach when it is also dilated, and be loose above and lightest along the lower iliac segment. The relief is often instantaneous and remarkable. A silk-and-wool knitted abdominal protector may be worn beneath it.

The special treatment of chemical dyspepsia is vested in remedies to regulate and supplement secretion. We possess few drugs which have a selective action on the pancreas. Ether is probably one of them, but its value on account of this property is more than counterbalanced by the harm it does in other ways. Pilocarpine in small doses is a remedy of some utility and power. But to increase pancreatic secretion we are forced to depend on constitutional remedies—massage, electricity, and nerve tonics. It is equally difficult to supplement the pancreatic juice. Pancreatin given by the mouth is either wholly or partly destroyed, partly absorbed, and partly passed on into the duodenum. If absorbed it is eliminated by the pancreas and liver, and in large doses may produce temporary diabetes by increasing the formation of hepatic sugar (Defresne). Clinical experience commends its administration under the protection of bicarbonate of sodium against the hydrochloric acid of the gastric juice.

Many remedies promote the flow of bile, but nearly all of them possess the disadvantage of interfering with gastric or duodenal digestion. Merck's salicin sweetens and tones the stomach and increases, but not to a very great degree, the flow of bile. It has not the inhibiting influence of salicylate of sodium on gastric and salol on duodenal digestion. It may, however, be necessary to administer a cholagogue, regardless of the temporary harm which it does. The administration of bile by the mouth has been highly praised by Dr. William H. Porter. Bile arrests artificial peptonization, but in the stomach exerts no disturbing influence on the chemical process, increases secretion, sharpens the appetite, and promotes nutrition (Dastre, Oddi). These are very strong statements, and are, of course, based on the introduction of a small quantity of bile into the stomach, from which it is absorbed to rapidly pass to the liver, the biliary salts thus gaining access to the entero-hepatic circulation. Bile is a digestive secretion, but an *excretion* as well. Nature and clinical experience seem to agree that it is well to keep it out of the stomach. A cholagogue is more apt to put some new, fresh bile into the duodenum, where it seems to belong. My limited experience with its administration by the mouth has been unsatisfactory.

To increase intestinal secretion, ipecac in small doses is a pretty reliable remedy. Large doses of an alkali may be required to supplement the alkaline carbonate of the intestinal juice.

To control gross symptoms we have all of the symptom drugs of the materia medica at our command. We should be careful to select such as do least harm to digestion. Antiseptics are popular, but do not seem to do much good. Cleanliness and regular peristaltic drainage are much better than antisepsis. Symptom drugs are rarely required if the remedies which impart systemic and local tone and strength, regulate or supplement secretion, and secure normal muscular movement are combined with a proper diet.

There is no other disorder of digestion in which the dietetic indications are so clear and so absolute. Intestinal errors are final, and occur right in the gateway of nutrition. A certain degree of freedom can be given the gastric dyspeptic, for the duodenum may correct the blunders or negligence of its assistant. But the diet of intestinal indigestion must be marked out in hard-and-fast lines. In the one a limited license may be tolerated; in the other the tyranny is unrelenting. In the one, concessions may result in a patched-up peace ; in the other, the rule is of iron. Additions to the diet may be cautiously and reluctantly made while the patient is under the eye of the physician, but in the beginning the control must be absolute and the firm grasp only slowly relaxed as the digestive ability of the intestine increases. I am now speaking of the cases in which there is an established defect of secretion or of motility, be it functional or organic, it matters not, so long as the capability of the digestive system is the dietetic guide.

The best diet in intestinal indigestion—and I state it with all the force of a wide experience—is a diet of lean meats. The worst foods are those that require the bile and intestinal juice to digest and absorb them. Intestinal dyspeptics digest incompletely and with the greatest difficulty sweets, fats, starches, and wines. We know that a good deal

of starch in some way disappears in the absence of
pancreatic juice, the steapsin only splits neutral fats
into fatty acids—and glycerin—while cane sugar
is inverted almost exclusively by the intestinal
juice. Milk occupies an intermediate position, because the intestinal juice has nothing to do with its
digestion. It is a popular error to suppose that this
mixed food is chiefly digested in the stomach. The
casein is divided by the lab-ferment of the stomach
into hemicasein-albumose, which is absorbed (with
or without further peptonization), and caseogen,
which unites with the alkaline earths to form cheese
and passes with the other ingredients on to the
duodenum (Arthus). In the beginning milk may
completely relieve the gastric symptoms, but the
objections to it are fatal. It does not give the duodenum rest ; it contains fat, lactose, and casein ;
an excessive quantity must be given to maintain
nutrition ; it cannot be employed when gastric dilatation is present as a complication. An exclusively milk diet is essentially a starvation cure
(Ewald). Whatever be the explanation, the phy-
siologist and chemical pathologist may decide. I
base my contention on clinical experience, and I
know that a diet of lean meats is the one most certain to give brilliant results. The diet may be ar-
ranged in three classes—the exclusive, rigid, and
advanced.

Exclusive Diet.—The lean meat of beef or mutton and the white meat of chicken. The muscle
pulp, free from fat and fibrous tissue, of the adult
animal only is permitted. The American chopper
in this country, and the Galante-Debove pulpifier

in France, are the best instruments. Skimmed meat juices. Whites of eggs cooked just enough to hold together. And to this list may be added Mosquera's beef meal. Lemon juice with or without horseradish. A cup of weak coffee or tea without sugar and cream, or a glass of hot water. This is the diet of the severest cases, and is soon supplemented by the articles of the second class.

Rigid Diet.—The articles of the exclusive diet. Broiled beefsteak or roast beef. Roast leg of mutton or broiled chop. White meat of fresh fish (sole, whiting, flounder). Soft part of raw, roasted, or broiled oysters. Cooked celery, watercress, crust of stale French roll. Dry toast with a little butter. Clear and unsweetened coffee or tea. A little diluted brandy or whiskey may be tried.

Advanced Diet.—To the preceding articles may be added broiled game, venison in season, sweetbread, eggs (poached), rice, cracked wheat, California wafers, wheatina—thoroughly cooked. Baked floury potato, French peas, string beans, tomatoes, and spinach (if no lithæmia). *Purées* of fresh vegetables. The juice of a few grapes. Milk warm from the cow or sterilized as soon as drawn. Tea or coffee without cream or sugar. Light claret or old dry sherry. A little Worcestershire sauce. No veal, lamb, hog meats, goose, duck, cod, herring, salmon, or other very firm and fat fish ; no old or raw vegetables, pastry, very acid or sweet fruits : no cheese.

This dietary is adapted alike to the chemical and motor varieties of dyspepsia, the varying element being the quantity of fluid taken with the meals.

The dry diet, first advocated by Chomel, is to be used in dilatation and deficient secretion. The five or six ounces of fluid should be slowly drunk after the meal, so that the stimulating action of the dry food on salivary and gastric secretion may be obtained. Starving these patients for fluid will not cure them; in the interval (which should be long) between meals enough water should be ordered to keep the urine in the proper condition, avoiding distention of the stomach and emptying it by the means already delineated. Hot water is rapidly absorbed and promotes downward peristalsis, increases primary oxidation and elimination, and is almost essential in the exclusive diet. In hyperchlorhydria water can be taken freely as a diluent and to prevent pyloric spasm against the passage of a hyperacid chyme.

Detailed and dogged supervision is the price of success. To prescribe a diet and then not to see that it is digested and assimilated is to court failure. By the right quantity and quality of food and water the urine should be kept free from deposit, of normal slight acidity, of specific gravity about 1.014 or 1.018, and without excess of coloring matter; the stools healthy, the patient without local distress related to eating and without abnormal flatulence, and the blood gathering hæmoglobin and red corpuscles. These are the clinical guides in the continued use of the systematic treatment.

Intestinal indigestion is not curable by drugs alone. The treatment must draw on a richer store of remedial powers. The much-drugged and neglected baby soon withers and falls away; the well-fed and carefully nursed child is of more vigorous growth. The one is a flower without roots and as weak as a life without good hygiene and the right foods. The very drugs, the warm sunshine which should be its strength, only hasten the approaching decay. Curative treatment is of a more vigorous growth, running down into the underlying systemic causes and twining its tender feeders about each unhealthy variation, and rising in its gathered strength, through physiological living, normal secretion and excretion, and careful alimentation, to a right performance of all the nutritive processes. We treat digestion, nutrition, and the nervous system, the physician and patient standing shoulder to shoulder in the struggle to bring the organism under the dominion of the gentle forces which make for health. The powers of evil that one cannot stay with iron chains the sweet influences of hope, contentment, and quietude will sometimes lightly bind.

CHAPTER IV.

THE CAUSATION AND TREATMENT OF CHRONIC DIARRHŒA.

It is not always possible to connect chronic diarrhœa with a distinct lesion of the intestine, nor can we limit its origin to functional or organic defect of the digestive system. It is by no means rare to find it one of the symptoms of disease of a distant organ, or disorder of nutrition, or defect of elimination. But chronic diarrhœa is a symptom so frequent that it may serve as a convenient point from which to begin investigation, so important as to often command our whole attention, and so predominant as to dictate the treatment. Whenever present it stands out in bold relief in the clinical picture and necessitates a search for its hidden meaning. The term " chronic diarrhœa " may be made to serve a useful clinical purpose, and no apology need be offered for selecting it as the subject of a paper based on investigations at the bedside.

Fluidity is the most constant characteristic of the diarrhœal stool. This physical quality results from excessive secretion or transudation, or increased peristalsis and diminished absorption. The stools are also altered chemically and microscopically, but the character of the discharges varies very much with the age, diet, the nature and location of the disturbance. The frequency of the evacuations is

also no criterion, and varies widely both in health
and in disease. But habit and other influences
establish a certain routine which, though it varies
with each individual, may be taken as a standard.
The character and frequency of the stools will
nearly always enable us to make out the presence
of the symptom. Moreover, diarrhœa, whether
conservative or not, is always an exhausting process, and when long continued must inevitably affect the general health. Hence chronic diarrhœa
may be defined as the frequent evacuation of the
fluid, and usually abnormal, contents of the intestine, with more or less impairment of the general
health.

A classification of chronic diarrhœa based on the
changes in the stools is not desirable. A careful
study of the stools will not fail to yield some useful
information. But in the same case the character of
the stools varies from day to day, and bears no definite relation to the lesions.

A classification based on etiology would be more
scientific, and stands in direct relation to the advanced treatment which strives to go beneath the
surface and strikes at causation. A careful review
of the possible causes will aid very much in formulating a rational treatment, but our knowledge at
present is too incomplete to enable us to make a
scientific etiological classification.

A nomenclature based on pathological findings
is neither desirable nor practical. Morbid anatomy is only a symptom of unhealthy cell activity, and widely different processes find expression
in the same tissue changes. But the lesions must

6

be taken into consideration in formulating the treatment.

In studying a case of chronic diarrhœa I constantly keep before me two objects of commanding importance: the detection of the proximate and remote causes, and the discovery of the nature and location of the intestinal lesion. In this way we gain all the information that is of most value in the management of the case.

The proximate cause of every diarrhœa is located in the intestinal wall. The intestine is a secreting, absorbing, and eliminating tube, which propels its contents in a peculiar way, and in which the most important part of digestion takes place. In diarrhœa too much fluid is poured out from the mucous membrane, or too little fluid is absorbed, or the contents of the intestine are hurried along too rapidly. It is common to find two, or even all, of these factors active in a particular case.

Diarrhœa from supersecretion is a frequent variety. It is commonly due to local irritation, with here and there patches of catarrhal inflammation. It is also found in chronic nerve or blood states, and may often be traced to auto-infection as the remote cause. Chronic dyspeptic diarrhœa may be taken as the type of this form. Much mucus and a disproportionate quantity of undigested food, especially starch, are found in the stools.

The excess of fluid may be an exudate, as in a condition of the mucous membrane analogous to an eczema or herpes. Or the fluid may be transuded in passive congestion, such as occurs in hepatic cirrhosis, obstructive disease of the lungs, or uncom-

pensated valvular disease of the heart. The pathognomonic sign of this variety is the presence of
serum albumin in the stools.

The intestinal mucous membrane is also an eliminating organ, and diarrhœa is not rarely due to
exaggeration of this function. The diarrhœa of
chronic Bright's disease and septicæmia are types
of this form.

Diminished absorption may be the starting point
of a diarrhœa. But a diarrhœa originating in this
way will not long remain simple, as the resulting
superdigestion, fermentation, and putrefaction will
produce supersecretion, exudation, and excessive
peristalsis. Impaired absorption always forms one
of the links in the ætiological chain, and has as much
to do with the persistence as with the causation of
diarrhœa. The stools contain completely digested
products.

Diarrhœa from excessive peristalsis is neuromuscular in origin, and occurs in its simplest form in
neurotics with lively reflexes or with hyperæsthetic
mucous membranes. Exaggerated peristalsis, however, usually results from a local irritant. A stool
occurs regularly and rapidly after each meal, and
consists chiefly of unaltered food.

From this it will appear that diarrhœa is wholly
or in part a conservative process in every variety,
except that which is purely nervous in origin, and
this variety, it must be admitted, is but rarely met
with.

These divisions are based on unhealthy variations
in the physiological processes—the surface-play of
concealed forces. While it will not clearly reveal,

the manner of appearance of the diarrhœa will suggest the salient features of the underlying disturbance. The kind of fruit or flower will enable us to infer something of the nature of the seed and the development of the plant. It is always difficult, and sometimes impossible, to discover the remote cause, be it located in a disorder of nutrition or hidden in the disease of a distant organ.

Disease of the kidneys, heart, liver, lungs, and spleen must usually be well marked in order to produce a diarrhœa. Anæmia, gout, leukæmia, Hodgkin's disease, scurvy, syphilis, tuberculosis, and septicæmia must also be passed in review and excluded.

A very large percentage of all cases of chronic diarrhœa find their origin in derangement of one of the three great processes of nutrition—digestion, absorption, and metabolism. The perfection of each one is essential to the integrity of the whole ; this constitutes the solidarity of the nutritive processes. From a therapeutic standpoint it is of great utility to locate the primary disturbance. The presence in the urine of the incompletely elaborated products of tissue waste, such as uric acid and the urates in excess, and of pathological urobilin, would point to faulty katabolism ; peptones, albumin, or sugar in the urine might implicate assimilation ; while the discovery in the stools of the digestive products in a fluid and diffusible form would suggest defective absorption. But the nutritive disorder more often takes its origin in the digestive tube, either in gastric dyspepsia or inflammation, with alteration either in the chemical process or in the muscular movements ;

or in intestinal indigestion from faulty chyme, bile,
pancreatic secretion, or intestinal peristalsis. An
insufficient diet, of which simple emaciation will be
the evidence; unhealthy food and impure drinking
water, an improperly constituted diet, will often be
found the initiating causes, though secretion and
muscular movement be in every way normal.

An important connecting link in the causation
of chronic diarrhœa is auto-infection, which may
be from the digestive or from the general system.
Absorption of the products of superdigestion, fermentation, and putrefaction is one source; defects
of assimilation and disassimilation, increased cell
activity and tissue waste, and incomplete elimination are others. The quantity of toxines formed in
health may be increased, or new ones may be manu-
factured in defective nutrition, or bacterial products be absorbed, and self-poisoning will result unless elimination is very rapid. Some of the toxines
dilate and others contract the blood vessels; some
alter the blood as well as the blood pressure, thus
impairing secretion or causing exudative or productive inflammation. Some paralyze, others excite,
the nerves; all exercise a pernicious influence on
nutrition. The prevention and treatment of auto-
infection is the most important part of the management of chronic diarrhœa.

It has been ably maintained that we never have
a diarrhœa without the presence of an enteritis.
But it is now a fact pretty well established by careful autopsies that diarrhœa frequently is not accompanied by noticeable lesions of the intestine.
From a practical point of view the detection of the

cause is of much greater utility than the diagnosis
and location of the lesion. The chief advantage of
a knowledge of the anatomical state of the mucous
membrane is the light it throws on prognosis. But
the nature and location of the lesion afford certain
indications in treatment.

I have been in the habit of grouping all my cases
into two large classes, according as there is or is
not a marked lesion of the intestine, and try to decide whether or not ulceration is present. This
classification is somewhat arbitrary, but it is usually possible to group the cases on this wide basis.
Our standpoint is at the bedside, and this broad
classification, in which minute anatomical distinc-
tions are not made, has a practical bearing.

In the functional disorder the symptoms are mild-
er and may be intermittent ; there are no persistent points of tenderness and no thickening of the
bowel, and the stools contain no products of inflammation.

A large number of chronic cases with intestinal
lesions follow acute attacks that have their remote
cause in the digestive system, or form part of the
clinical history of the acute infectious diseases.
The persistence of pain, tenderness, and fever would
indicate the presence of an important lesion. The
discovery in the stools of much epithelium, mucus,
and unaltered bile pigment, of pus, blood, false
membrane, and pieces of tissue from the intestinal
wall, would prove the trouble to be organic. Chronic
gastritis with chronic diarrhœa is accompanied by
chronic enteritis.

Ulceration of the intestine is simple, syphilitic,

tubercular, or malignant. The signs of a lesion, in many cases the strict localization and persistence of a painful and tender point, the presence and continuance of much pus, blood, and mucus without tenesmus, and the detection of bowel tissue in the stools, establish the diagnosis of ulceration. Intestinal carcinoma is usually located in the rectum, and can commonly be felt by the finger through the anus; cachexia and rapid decline will also point to malignancy. Simple ulceration results from a severe catarrh, or from acute or chronic follicular inflammation, and commonly involves a large extent of surface. Syphilitic and tubercular ulceration is more strictly localized, and the disturbance of the digestive process above the lesion is from excessive peristalsis. In syphilis we may get a specific history or characteristic skin lesion, or a persistent headache with periodical exacerbations and attended by insomnia and unwonted irritability of temper, an early endarteritis, or other sign of this protean malady. Tubercular ulceration is almost never found apart from pulmonary tuberculosis, and the rapid pulse, dry skin, hectic fever, and localized physical signs will confirm the suspicion. The absolutely pathognomonic sign is the discovery of the tubercle bacillus in a shred of the bowel tissue found in the stools.

The character of the stools, the persistent points of tenderness, and other physical signs, taken along with the clinical history, will locate the lesion with a good deal of exactness. The lower the lesion the more frequent and more painful are the movements. It is rare to find the small bowel alone dis-

eased. The large bowel is nearly always involved.
Commonly associated with it is disease of the lower
ileum. It is near the ileo-caecal valve that bacteria
abound, that fermentation and putrefaction are
most active, that irritants long remain in contact
with the mucous membrane. When either the
small intestine or the colon is alone diseased there
will be periodical attacks of diarrhœa ; when both
are involved the diarrhœa is likely to be continuous. Pain occurring just before a movement is
usually located in the colon. Tenesmus is present
only in proctitis. Indicanuria, fatty stools, recurring slight icterus, and persistent flatulence in the
small intestine are pathognomonic of duodenal defect. Much unaltered bile pigment and mucus intimately mixed with the fæces point to the small
intestine. When the trouble is located in the as-
cending colon the stools are soft, muco-feculent,
and little yellow globules of mucus are visible, and
hard fecal lumps coated with mucus from the
lower half of the large gut. When from the rectum the stools consist of yellowish or blood-stained
white-of-egg mucus or mucus and fibrin shreds ;
and the lower colon and rectum may furnish a
shred or cylinder, formed of a network of fibrin
filled with mucus, with here and there an epithelial
cell on the surface, or exfoliated casts of false membrane.

Having briefly reviewed such points in aetiology,
differential diagnosis, and localization as can be
utilized at the bedside, turn we now to the treatment.

Good hygienic surroundings, a regulated life, and

a proper diet will often suffice to cure a mild diarrhœa. But the severe cases must be subjected to a
rigid régime. Many of these patients have tried
everything, done nothing thoroughly, and lost faith
and hope. An important strategic point is already
gained if we win the confidence and arouse so
strong a desire to get well as to cause every energy
to be bent in the direction that we dictate. The
successful management of these cases depends as
much on the co-operation of the patient and the detailed observance of directions as on the skill of the
physician. It is not enough to order, but instructions must be carefully and cheerfully obeyed. Of
so great importance are co-operation and attention
to detail that I no longer try to cure these patients
against their expectation and will. They must acquire a soul-forwardness toward health—every
thought, feeling, and emotion must be enlisted in
the work.

Having secured the confidence and hearty co-operation of the patient, we give minute directions as
to clothing, bathing, rest, and exercise. From malnutrition and auto-infection the vaso-motor centres
are weak and irritable, and paling of the surface
leads to a corresponding internal congestion.
Hence the necessity for warm clothing, especially
over the abdomen, to protect against sudden changes or extremes of temperature and loss of body
heat. The rapidity and completeness of reaction
guide in the selection and the mode of bathing. In
the beginning a warm plunge or sponge bath in a
warm room should be advised, and the difference
between the temperature of the air and the water

cautiously increased from day to day. The bath
improves the function and nutrition of the skin and
tones the nervous system. It has been demonstrated that the toxicity of the urine is increased
during the administration of the Brand treatment
of typhoid fever, and the increased elimination of
toxines is not the least of the benefits derived from
bathing. If the stools are frequent and exhausting,
absolute rest in bed must be enjoined ; during con-
valescence moderate exercise and fresh air will hasten the cure. Overfatigue, mental and physical,
must be scrupulously avoided, temperance and
moderation being the guides of conduct. The mode
of life must be put on a physiological basis and as
much energy and vitality conserved as possible.

The curative treatment of a chronic diarrhœal
disease has very little to do with the control of the
symptom by the use of opiates and astringents ; we
must go behind the lesion of the mucous membrane
and strike boldly at causation. Behind the veil are
the hidden forces at work, beneath the surface are
the sources of evil. It is a waste of time to strike
at the shadow ; it is useless to close the volcano's
mouth while the subterranean fires are still burning. The curative treatment of a chronic diarrhœa
must be ætiological.

Active elimination by all of the emunctories is also
a sheet anchor in the treatment of chronic diarrhœa.
Free drainage is the first law of surgery, and free
drainage is a controlling principle in the treatment
of a chronic disease accompanied by or resulting
from auto-infection. We have already seen that a
chronic diarrhœa is largely a conservative process,

and is just as essential as the drainage of a septic
wound. Checking a chronic diarrhœa by astringents and drugs that paralyze muscular movement
before the digestive tube is made clean and sweet,
can only produce a violent explosion which will
widen old rents or find new points of exit where
resistance is weakest. So great is the danger of
auto-infection from the alimentary canal that Nature has well barricaded the system against invasion from this quarter. An active peristalsis to di-
vert the enemy, mesenteric glands and the liver to
arrest and destroy, oxidation to burn, the skin,
kidneys, and liver to turn aside or sweep away—
these are the strong barriers which our treatment
must support and strengthen. Impaired digestion,
defective absorption, malassimilation, auto-infection, are heavy blows against nutrition. To build
up the blood so that it may perform its work is
a controlling object. Healthy nutrition is a hope
that only careful alimentation can realize. These
are the important general considerations : on the
one hand the bright side of the shield, a well-fit-
ting armor, a determination to conquer, and on the
other the removal of the cause, careful alimentation, and active elimination.

Of no less importance are the local indications :
1. To cleanse the alimentary canal and keep its contents sweet. 2. To secure perfect digestion of the
food taken. 3. To promote absorption. 4. To diminish the work of the diseased part. 5. To treat
the lesions. 6. To treat the sequelæ. 7. To control the harmful symptoms.

Our first object is to cleanse the alimentary canal,

and cholagogues and purgatives will render efficient
service in its accomplishment. An increased flow
of healthy bile will meet more than one indication
—it is not irritating, is laxative, and also aids in
digestion, absorption, and the prevention of decomposition. Podophyllin, ipecac, salicylate of sodium
(or, better, salicin and bicarbonate of soda), and the
bichloride and biniodide of mercury are the most
useful cholagogues. To get their selective action
on the liver these drugs should be given in minute
doses. Small doses of calomel also act well, especially if the kidneys are sound, or the heart diseased, or arterial tension is high, or the bile ducts
distended. Cascara sagrada is the most valuable
laxative—it increases peristalsis by its action on the
nerve supply of the intestine, washes out the
glands and follicles by augmenting their secretion,
and in laxative doses is unirritating, an important
negative quality that often secures for it preference. Those drugs should be selected which least
irritate the diseased part : too much care cannot be
exercised in this respect, as these remedies cut both
ways and can do harm as well as good. Stomach-
washing will also help us to clean a part of the alimentary canal. When this important viscus is dilated and incapable of emptying itself completely,
when the muscular movement is defective and the
food is fermenting, decomposing, or undergoing
superdigestion, the procedure is a valuable one, but
must not be repeated too frequently. But when
not dilated, and strong enough to empty itself, the
stomach can be efficiently and agreeably washed
out by copious draughts of hot water. Hot water

is also a powerful hepatic stimulant, liquefies the
bile, and washes out the liver, which is often in-
fected from a septic duodenum or through the portal vein or hepatic artery. The liver is the great
central depot for the arrest, destruction, and elimination of toxic material, and the entero-hepatic
circulation should be frequently flushed out. Hot
water does this very rapidly and efficiently. The
large bowel is the seat *par excellence* of fermentation and putrefaction, and the most frequent source
of auto-infection. It can be thoroughly washed
out with warm or cold water, to which an alkali
should be added if there be much mucus in the
stools. The use of antifermentatives and antiseptics is rendered necessary by the inefficiency of
lavements, cholagogues, and laxatives to accomplish our purpose the cleansing of the digestive
tube. I use only a few of the drugs of this class,
the ones that I have found the most efficient—salicin, the biniodide of mercury, salol, and the subnitrate of bismuth. Salicin is the best sweetener of
the stomach, given in ten- to twenty-grain doses,
two hours after meals, or one hour before breakfast
and retiring. The biniodide of mercury is valuable
in small doses when the decomposition is in the
small bowel, chiefly on account of its action on the
liver. Salol is by far the best duodenal antiseptic.
These three drugs act locally, and also by exciting
a free flow of the natural intestinal antiseptic—
healthy bile. Cholagogues spur onward the entero-
hepatic circulation, as Rosenthal has shown that
both bile and the biliary salts are hepatic stimu-
lants. Calomel is also an antiseptic, and some aid

is derived from its passage along the intestine.
Subnitrate of bismuth reaches the large bowel, but
is not of much value unless given in very large
doses. These drugs are very useful in combating
putridity and maintaining the sweetness of the alimentary canal. It has been suggested that bacteria
have something to do with digestion; I gravely
suspect that enough will be left for this purpose
after we have exhausted our means in the efforts
to exterminate them.

It is also important to administer clean and sweet
food and pure drinking water. This is a matter of
more moment than the little attention we bestow
upon it would seem to indicate. How rapidly a
septic colitis subsides when an impure drinking
water is withdrawn! How great a change is sometimes wrought by forbidding a food that is too
"high" or has not been scientifically prepared?
Attention to little details like these sometimes
changes the whole course of the disease.

Having secured, as nearly as we can, a clean and
sweet state of the digestive tube, our next object is
to get perfect digestion of the food taken. This is
an aim second to no other in importance. Undigested food in the wrong part of the intestine is
an irritant. Rapid absorption is the chief barrier
against superdigestion, fermentation, and putrefaction, and perfect digestion is the essential preliminary to the easy and healthy performance of this
function of the mucous membrane. We attempt
to realize this high aim by a proper diet, and by
increasing or supplementing whatever digestive
juice we have reason to suspect is defective. If the

stomach is at fault in its chemical work we keep
our eye on the acidity of the secretion, for the HCl
is an important and the most frequently varying
constituent of the gastric juice. The dilute HCl
should be given in two or three doses of five or ten
drops each, within two hours following the meal,
and a small quantity of fresh pepsin may be added.
I suspect that a dose of toxines is often given in
the name of this ferment. In the meantime we
give such drugs as are known to increase or diminish the acidity of the gastric juice. If the liver or
pancreas be at fault we use the drugs that have a
selective action on these glands, and supplement
with fresh bile and pancreatin by the mouth. It
is best to precede their administration by an alkali.
The time of giving them is two and a half or three
hours after meals, except on the milk diet, when
the proper time is just before each feeding. Duodenal digestion is thus made to begin in the stomach. If the muscular movements of the stomach
and intestines are defective, strychnia, massage,
and electricity will render important aid. Diarrhœa not infrequently has its cause in localized defective peristalsis—the contents collecting in the
weak and dilated parts and undergoing putrefaction, fermentation, and hardening. With a clean
digestive tube, the secretions and movements of
which have been regulated and supplemented, it
remains to select a proper diet. This is the most
difficult and most important part of the treatment.
And here the physician should dismount from his
"hobbies" and renounce so-called "fads" and
"cure-alls." Vegetarianism will rarely fail to do a

good deal of harm ; the milk diet in its many forms
is not a panacea ; a diet of animal food will not
often fail to benefit, and has a very wide range of
usefulness.

In selecting a diet we have a good many things
to take into consideration. The evils of exclusiveness all are ready to admit. In any dietary the
primary principles must be made to preserve a certain proportion in obedience to the laws of physiological chemistry, and such proportion arbitrarily
altered to suit the needs and capabilities of general
nutrition. But laboratory results need to be corrected and controlled by the testimony of the human digestive system. The diet habits of mankind
and of the different nations of the earth furnish
a rich store of information ; for man, when permitted to do so, eats what most pleases the palate,
keeps him well nourished and strong, and gives the
least after-pain. Climate, age, activity, peculiarities, and the capability of the digestive organs are
other important considerations. Now, in the diet
of a chronic diarrhœa the food must be chiefly
digested by the stomach, contain the right proportion and proper quantity of proximate principles
to meet the requirements of secretion, nutrition,
and the production of animal heat, and leave no irritating or indigestible residue. Denutrition must
be guarded against, and the diseased intestine given
physiological rest and kept free from irritation. An
exclusive diet of milk, or a diet of meat free from
fibrous tissue, would fulfil these indications—the
one more completely than the other, perhaps—but
both must be perfectly digested. Milk is a fluid,

but becomes semi-solid during digestion. Meat is a
solid, but becomes a fluid in its preparation for absorption. Milk may be a little more easily assimilated, but, bulk for bulk, is not so nutritious. The
final product of the perfect digestion of the one is
about as easily absorbed and unirritating as that of
the other. Both require great care in selection,
and the meat must be properly prepared and
cooked. However, it is difficult to get, day after
day, milk which is free from pathogenic bacteria :
it readily undergoes, both in and out of the stomach,
chemical and bacterial changes with the formation of irritating and poisonous products ; and I
have found it well-nigh impossible to secure its continued perfect digestion during a period long enough
for a cure to take place. When the gastric juice is
hyperacid, or duodenal catarrh or portal congestion
or excessive fermentation is present, milk will not
agree. The supreme test is the one at the bedside.
In my experience a meat diet is much more valuable, less dangerous, and of a much wider range of
application than milk in the treatment of chronic
diarrhœa. Exclusive in the beginning, the meat
must be supplemented by bread, cereals, and the
more easily digested vegetables in the manner detailed by me in a paper printed elsewhere in this
book.[1] It is the duty of the physician to see that
whatever food be taken is completely digested and
assimilated, and he has in the daily physical examination of the digestive system, and the analysis of
the urine and the inspection of the stools as often

[1] See clinical paper " On the Treatment of Functional and Catarrhal Diseases of the Stomach and Bowels," Appendix, p. 113.

as may be necessary (aided, if needed, by the microscope), a pretty sure guide. If the patient feels no
pain nor discomfort nor drowsiness after meals, if
there is no flatulence, if the urine contains no abnormal coloring matter nor excess of phosphates,
urates, or uric acid, and the stools contain no undigested products, we know that the food is being
digested and assimilated, and, if there be no loss of
strength, absorbed in sufficient quantity to meet the
demands of life.

To avoid denutrition is not alone requisite : the
barriers must be made strong ; the body must be
protected and defended and built up. Not only a
pure and adequate but also a rich blood is needed.
And the quality of the blood, its gain or loss of
richness from day to day, can be detected by count-
ing the corpuscles and measuring the hæmoglobin.
No physician would now assume the management
of a disease of the heart or lungs without the evidence and guidance of physical signs. No physician should now attempt to diagnosticate or treat a
disease of nutrition without a study of the blood
and excretions.

When we have an alimentary canal clean and
sweet, and the lining washed free from mucus, containing a completely digested and unirritating fluid,
much has already been done to promote absorption.
An active entero-hepatic circulation and the control
of excessive peristalsis should complete the work.
The relief of portal engorgement and the stimulating of the liver will aid the one, while the removal of local irritation and the quieting of the
nerve endings and centres, and the strengthening

of them by active elimination and improved nutrition, have done much to realize the other. These are the curative means, but it is often necessary to control excessive peristalsis in order to keep the contents in contact with the mucous membrane long enough for absorption. Antacids, bismuth, and antispasmodics should be used instead of opium and narcotics. The control of flatulence also increases the absorptive surface.

The value of rest in the treatment of a disordered or inflamed part cannot be overestimated. Repair is more complete, healing goes on more rapidly. An exudation in the process of organization is easily broken up by movement. Absolute rest of a diseased intestine cannot be attained without stopping drainage, but a great deal can be done by keeping the part free from irritants, and by the use of drugs that will lessen the exaggerated irritability, that will quiet the pathological unrest. The diet should also be selected so as to diminish the work of the diseased part. When the duodenum is the centre of disturbance (as it often is) the stomach must be made to do the work. When the disease is lower down the diet must be such as is quickly digested and rapidly absorbed, and excessive peristalsis controlled. When the stomach and duodenum are able to do their work well, and the disease is in the colon only or low down in the ileum, a milk diet, if it agree, is superior to any other.

In the severe cases of chronic diarrhœa, when the muscular layer is atrophied or œdematous, or infiltrated with inflammatory products, constipation is very apt to supervene as soon as irritation is removed. I would like to emphasize this important clinical fact—that these weak points in the intestinal wall are often localized, and the obstruction in the drain must be overcome by massage, laxatives, and lavements. To clear out these depots of fermentation and putrefaction is an essential part of the treatment : until this is done there can be no rest, no healing.

The indications afforded by the lesions have been partly met by cleanliness, rest, and the prevention of irritation. If the lesion be syphilitic, specific treatment must not be neglected. When situated in the large bowel something may be accomplished by medicated lavements.

The treatment of the sequelae resolves itself into the treatment of atrophy and deformity—the results of degeneration and destructive inflammation. A proper diet and a regulated life will aid Nature in the readjustment of the organism to the changed conditions. The deformity may demand the surgeon's skill.

The special and general treatment of chronic diarrhoea must often be modified or supplemented by the treatment of the causative disease.

In conclusion, the indications for the treatment of chronic diarrhoea may be thus briefly stated : 1. To remove or treat the cause, which presupposes its detection. 2. To improve nutrition and conserve energy. 3. To secure active elimination and prevent auto-infection. 4. To cleanse the alimentary canal and keep its contents sweet. 5. To secure perfect digestion of the food taken. 6. To promote absorption. 7. To diminish the work of the dis-

eased part. 8. To treat the lesions. 9. To treat
the sequelae. 10. To control the harmful symptoms.

A broad and comprehensive and ætiological treatment, and one which I have found successful—a
union of many powers which make for health, a
union in which "all are needed by each one." It
is not sufficient to meet the controlling indications,
but regulations must descend into minute details.
The moral management of the patient has a powerful and practical bearing. Two important elements
of success are individualization and the persistent
doggedness with which one enforces right living.
The prescription of drugs is a very small part of the
work which we have to do. The chief aim, the
definite therapeutic purpose, is to secure healthy
nutrition by careful alimentation, perfect digestion,
and complete elimination, thus keeping in active
circulation a pure and rich nutritive fluid. In no
other way can we control and strengthen cell life
than by placing it in the best environment and ob-
taining the substitution of new protoplasm for that
which is old and diseased. This is the basis of cure,
the grand purpose which gives unity and system to
the management.

CHAPTER V.

HABITUAL constipation, as it will be considered in
this short chapter, may be defined as chronic inadequate intestinal peristalsis. The defect is a purely
neuromuscular one, and must be carefully differentiated from cases in which there is more or less
stasis and retention of the intestinal contents from
obstruction. Here it is not inefficient peristalsis,
but the obstruction, whatever be its nature, that is
the disease.

Peristalsis is normally under the control of the
nervous system through the reflex stimulus of the
intestinal contents, and consequently there are three
ways in which the disorder may be produced—by
defect on the part of the nervous system, or of the
muscular layer, or of the peripheral excitation of
the sensory nerves of the mucous membrane. Now,
the one fact, on which a good deal of what follows
will be based, is that the normal stimulus of intestinal peristalsis is the unabsorbed product of healthy
digestion, and, consequently, when there is no primary neuromuscular defect we must look for the
origin of the trouble in indigestion, defective secretion, or in the quantity or quality of the food and
drinks.

The physical properties of the intestinal contents
depend on the nature of the diet, the quantity of
fluid swallowed, and the rapidity of absorption and
elimination. In polyuria, and when too little water
is drunk, the faeces quickly become hard and dry.
Absorption from the stomach and the duodenum is
not very great as compared with its activity lower
down in the small intestine and in the colon. A
diet containing a large quantity of indigestible matter will prove to be mechanically too irritating. An
abuse of starches is the most common cause of disordered peristalsis depending on the nature of the
diet. It has been demonstrated on a grand scale
that the starchy army diet produced either diarrhœa
or constipation. And it has been conclusively
proven that when fatigue, irregular habits, and
unsanitary surroundings are excluded, a diet of
starches, in healthy men, causes constipation and
diarrhœa. Severe irritation sets up diarrhœa.
Mild, long-continued irritation will just as surely
establish constipation. The mucous membrane becomes too tolerant.

Habitual constipation may be either the cause or
the result of disordered digestion. We have already
seen in the preceding chapters how intimately associated are the chemical and motor functions of
the digestive tube. The digestive changes that the
food undergoes are about finished at the ileo-cæcal
valve. The chemical alteration of the food mass in
the colon is chiefly due to organic fermentation and
decomposition, and, by a beautiful provision, nature
has made these decomposition products (scatol,
H_2S and CO_2) the active exciters of peristalsis.

But, unfortunately, these substances are more or less
poisonous, and when not expelled undergo absorption along with abnormal products, and copræmia
with its restlessness, giddiness, insomnia, pains and
mental depression, anæmia, chlorosis, palpitation,
cold hands and feet, and digestive disturbances,
results. Auto-infection disorders digestion, deranges the nervous system, and lowers nutrition.
It is thus that the vicious circle is established and
continues its unceasing revolutions. The constipation results from the diminished sensibility which
follows the chronic irritation or inflammation or
distention produced by the imperfectly digested and
decomposing and fermenting food mass. In the
same way constipation originates in the abuse of
purgatives and neglect of the normal promptings
of nature. It is the pill-taking American, and mod-
est woman, and busy or lazy or negligent man who
most often contract the habit in this way. When
the call to stool is unanswered the fæcal matter is
either regurgitated by reversed peristalsis into the
sigmoid flexure, or accumulates unheeded in the
tolerant rectum to undergo hardening by absorption. Thus is the unhealthy variation established
by bad habits and unphysiological living.

We have already considered the relation of neurasthenia to the neuromuscular form of dyspepsia.
Peristalsis and tonicity are inadequate, because
too little nerve power is radiated out to the muscular system. There is a lack of muscular power,
and a lack of muscular tone develops from dis-
use. It matters not what may be the disease of
which the neurasthenia is the symptom, or the

nature of the cause—emotive shock, overwork,
traumatism, or malnutrition—of which it is the
result. The lowered nerve tone, the nerve weakness, is the cause of the diminished vitality and
denutrition of the muscular layer and the inadequate peristaltic power. The neuromuscular insufficiency is so often associated in families as to
suggest the influence of heredity ; but, while not
prepared to deny the possibility of the inheritance
of the specialized defect of constitution—it being
well known that unhealthy variations are transmitted with the same certainty as are the useful
ones—it seems more plausible to suspect that the
vice which arrogance is wont to attribute to the
sins of another is nearly always acquired by bad
habits, a faulty environment, and unhealthy living.
Infectious and mineral poisons like lead seem to
produce constipation by their influence on the nerve
supply. Chronic diseases of the brain and spinal
cord are also accompanied by obstinate constipation.

The cerebro-spinal and ganglionic nerves may be
efficient in the performance of their work, and constipation result from atony, or degeneration, or
atrophy, or œdema of the muscular layer. Here
the disorder has a muscular and not a neural basis.
A weak diaphragm and flaccid abdominal wall and
general muscular flabbiness are commonly associated with the atonicity of the muscular layer. The
inactive centres of old age go along with the athe-
roma and fatty degeneration and weak involun-
tary muscles. But more frequently the muscular
inadequacy is the accompaniment or legacy of a

diseased mucous membrane, peritonitis, or malnutrition from the distention of gases, or the pressure
of accumulated and hardened faeces ; or the œdema
of heart disease, or portal obstruction, or Bright's
disease, or of a watery blood.

Habitual constipation is without urgent distress :
it is slow in its destructive work and insidious in
undermining the general health. But intestinal obstruction is not rarely engrafted on habitual constipation, and whenever it supervenes the symptoms at once become severe. The condition is no
longer simply a disturbing but a deadly one. It
becomes, then, our duty, before a prognosis can be
given and a rational treatment adopted, to differentiate chronic inadequate intestinal peristalsis and
chronic constipation accompanying other diseases
and conditions ; to differentiate faecal impaction sequential to habitual constipation and faecal impaction or stasis due to the intestinal paralysis of peritonitis, the cæcal paresis of appendicitis, and complete obstruction produced by other causes.

Chronic constipation is a frequent symptom of a
diseased rectum or anus. It is advisable, in search-
ing for the cause of the constipation with a view to
arriving at a correct diagnosis on which to base an
opinion and palliative or curative treatment, to
make a careful rectal examination. When pain
accompanies and follows defecation this examination is imperative. An ulcer, or a fissure, or a blind
or complete fistula, or a sensitive pile, or an irritable
and powerful sphincter, will frequently be found
the disease which demands treatment. The fissure
or ulcer or haemorrhoid may be the result of the

constipation, in which case the neuromuscular disorder will persist after the cure of the local trouble. An eczema in the region of the anus (frequent in infancy) becomes a common cause of constipation through voluntary or reflex inhibition of defecation. The little child strives to prevent the suffering associated with the act. It is through frequent voluntary resistance that the sphincter is overdeveloped and the rectum made tolerant. Excessive hypertrophy of the body of the uterus, or a retroverted or retroflexed uterus, may be another cause of constipation.

Chronic intestinal obstruction must be established as the cause of the chronic constipation by the sequence of symptoms as revealed in the clinical history, by the detection of the causative lesion, and by the presence of additional symptoms to those ordinarily produced by habitual constipation. Very large and foul movements should excite suspicion. Habitual constipation is temporarily and painlessly relieved by the proper dose of a purgative, which would excite colicky pains above the site of obstruction. The mode of origin is of more importance than the symptoms. Previous severe inflammation would suggest bands or adhesions or constricting organized fibrous tissue. Ulceration is a common cause of stricture. Acute intussusception, ending in recovery by the formation of adhesions and the separation and discharge of the incarcerated part of the bowel, may be followed by chronic obstruction. The intestine just above the obstructed point hypertrophies ; peristalsis and thickening may be seen and felt ; dilatation may

alter the configuration of the abdomen. The form
of the faecal discharge may be important if piles
are absent, or the prostate is not enlarged, or the
uterus is movable and in its normal position. The
trouble may be revealed by the finger or the rectal bougie, or by filling the colon with water or by
inflating it with air. It is not always possible to
form a definite conclusion after the most careful
and exhaustive study.

Obstruction by faecal impaction, or the complete
and insuperable stasis of the intestinal contents, as
a sequence of habital constipation, is usually located
in one of the flexures of the colon or in the caecum.
It is more frequent in women. A history of long-
continued constipation becoming more and more
obstinate, the slight tenderness over a faecal tumor
which can be felt and indented, are the diagnostic
signs. The normal temperature, the marked abdominal distention without tenderness, the late occur-
rence of vomiting which is almost never faecal, the
extreme foulness of the breath, the increasing rapidity of the pulse and the gradual exhaustion by
chronic shock and inanition, and the fact that the
acute symptoms followed the administration of a
purgative, aid in the differentiation from the impaction of mechanical obstruction as well as the impaction produced by local intestinal paralysis.

The cardinal symptoms of obstruction and strangulation are the same—abdominal pain, vomiting,
and obstinate constipation; but strangulation is
acute, the onset is sudden without premonitory
signs, collapse is early, an external strangulated
hernia may be detected or a history of abdominal

injury obtained, a little bloody serum and mucus may be passed, and the urine contains albumin rather than indican.

Faecal impaction located in the caecum is both a cause and a result of appendicitis. Primary appendicitis and peri-appendicitis do not seem much more frequent than primary salpingitis and local peritonitis. The analogy between tubal disease and disease of the appendix is close enough to be instructive. Perityphlitis and abscess are about as rare as pelvic cellulitis and pelvic abscess. Typhlitis, on close study, will not be found much less frequent than endometritis. Pelvic peritonitis without tubal disease is as rare as localized peritonitis in the right iliac fossa that is not caused by a diseased appendix. In both we get closure or obstruction of the mouth, and accumulation of the secretions, and tubal or appendicular colic. The lumen of either tube may be the site of stricture. Sepsis may extend from the endometrium or from the mucous lining of the caecum. Purulent inflammation may travel in the same way. Pyosalpinx has its analogue in the accumulation of pus in the appendix. Chronic recurrent appendicitis is as difficult to cure without removal as chronic catarrhal or productive salpingitis. The analogy serves a useful purpose in emphasizing the ætiological relation of fæcal impaction of the caecum and typhlitis to appendicitis. Dilatation or distention of the caecum may also open the mouth of the appendix and permit foreign bodies, which may become incarcerated and produce ulceration and perforation or gangrene, to enter. The cæcal paresis and fæcal accumulation associated with ap-

pendicitis, and produced reflexly or by contiguity of
the inflammation in and around the appendix, is accompanied by fever. When the appendix is sound
fever is usually absent, since perityphlitis and local
peritonitis are so rare without appendicitis as to be
almost excluded from consideration. The differentiation of the cæcal accumulation sequential to hab
itual constipation and producing appendicitis, from
the accumulation of fæces in the cæcum resulting
from peri-appendicitis, cannot be made in the dim
light turned on by the clinical history and the physical signs.

The curative treatment of habitual constipation
is comprised in four special indications of commanding importance :

1. To secure perfect digestion.
2. To tone the nervous system.
3. To strengthen the muscular layer.
4. To attend to the hygiene of defecation.

To correct the special defect, to establish normal
and adequate peristalsis as the habit of life by
obeying the laws that condition it, is to make the
basis and purpose of treatment rational and curative.

A popular way of curing habitual constipation is
to prescribe an indigestible diet and force it through
with a purgative. Such treatment is irrational,
harmful, and never cures, but produces a temporary and deceptive improvement. Excessive irritation need not be expected to yield a very brilliant
result when chronic irritation has been the cause of
the trouble. The worst kind of a laxative is undigested food undergoing organic fermentation and

decomposition, and it does not seem to be a very good plan to derange the stomach and duodenum in order to make the colon empty itself. Artificial indigestion is not a cure for habitual constipation.

The needs of general nutrition and the capability of the digestive organs are the guides in the selection of the diet. The method of securing perfect digestion has already been fully discussed. The quantity of fæces passed varies with the nature of the diet, the completeness of digestion, the activity of absorption and secretion, and the rapidity of peristalsis. When digestion and absorption are good and the food is digestible, the patient must not resort to the pill box, because a stool that analysis proves to be normal in composition is small.

The drinking of too little fluid is a common cause of habitual constipation. Purgative and laxative mineral waters are constipating in their after-effects. Cold water increases tonicity. Hot water, as is well-known, is an active exciter of peristalsis. The urine should be kept at about 1.014 specific gravity, and the stools soft by abundant drinks and active intestinal secretion.

The constitutional measures and drugs for the improvement of secretion and motility have already been considered. To tone and strengthen the neuro-muscular element, massage, electricity, and strychnine are the most useful remedies. The accessory muscles of defecation should also receive attention.

The healthy stimulus of a normal digestive product and adequate neuromuscular power should be supplemented by regular habits. The unhealthy variation often originates in negligence, voluntary resistance, and irregularity. Frequent infraction of the laws of health is an influential factor in the causation of chronic disorders ; physiological living is a powerful remedy in their cure.

The materia medica supplies us with two drugs which, when rightly used, exert a curative influence in habitual constipation—aloin and cascara sagrada. Purgative doses do only harm. Aloes in large doses produces griping pains, congests all the pelvic viscera, and causes albuminuria. In small doses it is tonic, a mild cholagogue, non-irritating, and increases secretion and peristalsis. Its valuable selective and stimulant action on the muscular layer of the colon and rectum, without irritating the mucosa, makes it a valuable curative drug. It is not followed by constipation, and its long-continued administration does not lead to the formation of a pill habit. Aloin, one-tenth to one-fifth grain, is better than the crude drug and may be combined with ipecac, nux vomica, or a bitter tonic, as may be indicated.

Cascara sagrada is a valuable laxative with curative properties. It tones and increases peristalsis and intestinal secretion, and is a general tonic with a selective influence on the sympathetic system. The curative properties are also only manifested when given in small doses short of a laxative effect.

Purgatives, injections, suppositories of glycerin, etc., and other symptomatic remedies, do not come up for consideration in the curative treatment, which is comprised in good digestion, the hygiene of defecation, physiological living, and the strengthening of the neuromuscular layer.

APPENDIX.

A CLINICAL PAPER ON THE TREATMENT OF FUNCTIONAL AND CATARRHAL DISEASES OF THE STOMACH AND BOWELS.

THE purpose of this paper is to present the essential features of certain methods of treatment which I have found to be very useful. I shall endeavor to state them in a distinctly clinical manner, so as to show their practical application. It will be impossible to so enunciate them as to fit all cases, but I hope to convey an idea of measures that can be adjusted as they are needed. There is such a vast range between a functional derangement and an old chronic gastro-intestinal catarrh that no system of set rules can be made for uniform application. Accordingly, much will remain to be done in the way of wise adaptation by the good sense and skill of the physician. It is hoped, however, that the methods here outlined will prove to be generally applicable and of great service in one of the widest fields of practical work.

The first step in the treatment of functional and catarrhal diseases of· the stomach and bowels. whether moderate or severe. is to obtain full con-

8

trol of the mental condition of the patient. Too
much emphasis cannot be put on this point, as failure here is sure to mean failure in the future. If
there is a mental antagonism on the part of the patient to what the physician is attempting to do ; if
there is a lack of faith and willing co-operation ; if
there is, from first to last, a sort of secret indifference and resistance –then all treatment, no matter
how judiciously advised and how worthily applied,
is almost sure to result in failure. On the contrary,
if the physician first obtains the confidence and respect of his patient, secures his cheerful submission
to all instructions and requirements, and has his
glad and hearty endeavor to help bring about a
cure, then the principles and methods I am about to
offer are almost absolutely sure to result favorably,
even in the worst of cases. All this can be done by
patience and tact, and it is of first importance because of that close relation existing between the
brain, the sympathetic nervous system, and the organs of digestion. The influence of the mind over
the body is simply tremendous, and both the patient
and physician need to have such a great vital force
working with determinate action toward health.

The second step in the treatment of these cases is
no less important than the first. It consists in thoroughness, and repeated thoroughness, in examining
into the patient's condition. It is in the highest degree essential to interrogate over and over again
every organ, and to find out just how it is doing its
work. To obtain the desired information there is
no better method than carefully inquiring into the
patient's habits of life, the duration and severity of

his subjective symptoms, the significance of every physical sign, and then supplementing all this with a microscopical and chemical study of the blood, the urine, and the fæcal discharges. Careful, scientific study of the products of the system, made daily, is eminently important and useful. The reason lies in the fact that, if the machinery of the system is out of order, its products will be faulty ; and hence, by studying abnormal products, one is enabled to read, as it were, the condition of the organs that made them so. If a study of the urine reveals the state of the kidneys, is it not just as reasonable to believe that a microscopic study of the blood and fæces will disclose the state of the stomach, bowels, and blood-making organs? It would seem to need no argument, then, to prove that a daily thorough investigation of the excretions and products of the system is in the highest degree useful. It is the only means of accurately determining to what extent patients are digesting their foods, to what degree the liver and kidneys are doing their work, and just what quality of blood is being made.

As a third preliminary consideration it is highly necessary to place every patient under the most favoring hygienic conditions. In those cases where the affection is slight or limited it may not be necessary to impose more than a few reasonable restrictions upon diet, habits of life, and hours of work and rest. On the contrary, where there is very much catarrhal disease of either the stomach or bowels, it is usually necessary to confine the patient to his home for a time, and carefully regulate his work, recreation, diet, and medical treatment.

Indeed, everything pertaining to habits of business and life should be so regulated as to save nerve power, and the severer the disease the greater the necessity of this. To insure this result in bad cases the patient should rest half an hour before meals and an hour and a half after meals. In other and still severer cases it is better to insist on the patient's resting, sleeping, if possible, from one to two hours every forenoon, or else on his not getting up until an hour and a half after breakfast, and retiring immediately after lunch and remaining in bed until the next morning. The great object and end is to so regulate the life of the patient as to avoid "overwork and underrest," economize nerve force, and acquire a quiet, calm, tranquil state of body and mind.

Having thus first gained the confidence and good-will of the patient and directed him in regard to his habits of life, his diet and rest, the next thing is to endeavor to remove from his stomach and intestines, and also from the kidneys and liver, all morbid material. As you are well aware, the lining of the stomach and bowels in the diseases under consideration becomes coated, as it were, with the morbid products of supersecretion and fermentation. The secretions, being in excess for a long time, become thick, tough, and stringy. They are highly acid and laden with the germs of fermentation. Moreover, as a rule, the liver and kidneys are in an abnormal state and burdened with an immense amount of morbid material. All these vitiated and unhealthy accumulations need to be eliminated from the system. In other words, the surfaces of the alimentary

tract need to be washed off and the organs flushed
out in order to put them in a healthy condition.
Especially is it necessary to remove the bile from
the blood and stomach. Every one knows that the
effect of a large amount of bile in the stomach of a
well person is to greatly interfere with the appetite
and with the stomach digestion. If such is its effect
in people who are otherwise well, it is not difficult
to imagine what its presence does in the stomachs
of those who are in poor health and suffering from
catarrhal disease of the stomach. Hence the importance of freeing the stomach of vitiated, offensive mucus and bile by giving to it a rapid downward
action. This can be done in several ways, but I
know of none so simple, so grateful, and so effective
as washing it out by drinking hot water. Long experience has now shown that quantities of hot water
dissolve and liquefy the mucus and bile, stimulate
the secretory and excretory glands, and excite
downward peristalsis of the bowels. It is believed
that morbid substances are rapidly eliminated from
the system in some such manner. And this leads
me to say that in such cases hot water needs to be
taken systematically, under the direction of a physician who appreciates its utility and knows what
effect is to be achieved. At the beginning of treatment it is a good rule to order the patient to take
one glassful an hour or an hour and a half before
each meal and on retiring, increasing or decreasing
the quantity according to the rule to be given further along. It should not be taken too hot, but about
as hot as after-dinner coffee, or at a temperature of
from 110° to 120° F. The patient should be charged

to take it very slowly, consuming fifteen or twenty
minutes in sipping a glassful, in order to avoid scald-
ing the mucous surface of the throat and stomach.
Water taken too hot may injure the lining of the
stomach, produce a dry, feverish condition, or act
too powerfully and promptly on the skin. There
are other precautions to observe, which I will mention. If the glassful or more taken at bedtime
causes too frequent urination during the night, it can
be dispensed with; if the patient has a weak heart,
large quantities of hot water should be taken very
slowly ; if the patient has a tendency to hæmor-
rhages, the water taken should not be much more
than lukewarm and should be taken very slowly :
and if the patient is a woman subject to long-continued or excessive menstruation, she, too, should
take water very slowly and at a low temperature.
These precautions need to be observed so as to avoid
ill effects and dangers that might otherwise supervene. If at any time the hot water is disagreeable
to the patient, a little salt, pepper, lemon juice,
aromatic spirits of ammonia, or any innocent flavor-
ing extract may be added to suit the taste. If hot
water seems to nauseate the patient, its use should
still be persisted in, since this is a positive evidence
that the stomach is in a foul condition and needs
cleansing ; and, as evidence that cleansing does take
place, it can be said that, after an abundance of hot
water has been used for a time and the bowels get
to acting from two to four times daily, as they frequently do, the discharges are often either black and
sticky, or granular like coffee-grounds, or else they
contain masses of exfoliated, gelatinous mucus.

We often hear it said that the free and prolonged use of hot water tends to injure the system. Some say that it is weakening, that it weakens the nerves of the stomach, that it causes anæmia of the stomach, that it interferes with digestion, that it tends to produce a flushed face and cerebral hyperæmia, that it debilitates the alimentary tract, and that it causes a host more of most direful evils. As a rule, all these objections are theoretical and come from those who never used it intelligently and systematically, and hence are ignorant of the facts. In reply to such objections, all I can take time to say is that I have used hot water daily for six years without the slightest perceptible injury, and have seen only uniformly good results in persons for whom I have prescribed its daily and long-continued use.

As all are aware from experience, it is always a difficult problem to successfully feed patients who are suffering from diseases of the stomach and bowels. There has ever been a demand for some article of food that would not ferment, that would afford a maximum amount of nourishment, and that would be promptly and easily digested. At last such a food has been found, for we know that an animal diet, or, to speak more specifically, good, well-prepared muscle pulp of beef, can be relied upon for the purpose before named. Inasmuch, however, as beef varies greatly in its quality, it is necessary to exercise care in selecting that which is best, and this is found in the centre of the round of a well-fatted, corn-fed animal from three to six years old. This portion is freest from fat and is

the richest in those nutritive elements required by
the human system. It should be given to the patient in the form of beef pulp, which may be prepared by the process of scraping, or by passing it
through a "chopper" made for the purpose. The
object of such preparation is to remove all of the
fibre and leave the pulp in a condition to be both
palatable and easily digested. When the fat and
fibre are entirely removed, the pulp can be made
into cakes containing the number of ounces the patient is able to digest. These cakes should be from
half to three-quarters of an inch in thickness, care
being taken not to pack them too firmly. The
cake of beef pulp is then to be broiled over a slow
fire, preferably charcoal, until it is so cooked that
the outside is of about the color of ordinary broiled
steak and the inside of a pinkish hue. Great care
should always be taken not to overcook the beef
cake and so make it dry, brown, juiceless, and indigestible. If it is cooked just right, patients will
not tire of it, it is more easily and thoroughly digested, and all dangers from tapeworm are avoided.
In rare instances beef prepared thus is not palatable
at first, and when such is the case it can be broiled
between two pieces of dried or chipped beef, or a
few oysters may be broiled with it so as to impart
their flavor, or a few spoonfuls of beef blood or ex-
pressed beef juice freshly extracted from the beef
may be added. The effort should be to employ
simple means to make the beef palatable to the
peculiar tastes and fancies of the patient. The
beef pulp thus prepared should be given in small
quantities at first, not over four or six ounces in a

day, until its effects have been carefully noted.
Later on, as the stomach and bowels become
cleansed and more tolerant, the quantity may be
increased to eight, ten, twelve, fourteen, or sixteen
ounces at a meal. If patients tire of beef prepared
in this manner, or if it is very distasteful to them,
it is better not to insist on their taking it for a time,
but to let them have instead a lean chop, or a small
plain steak, or a little game of some kind, like
broiled grouse or pheasant. This change, however,
should be as temporary as possible, and an early return made to beef pulp, for from this comes the
maximum nourishment from the minimum effort.
If the functional or catarrhal condition is not too
severe, a limited quantity of starchy food may be
given, such as a small piece of stale roll, or a piece
of dry toast about one or two inches square. It
sometimes happens in these cases, and under this
restricted animal diet, that the patient's appetite
will seem to fail. When such is the case it is in-
variably due to either bile in the stomach, to undigested food, to a tired and depressed state of the
nervous system, or else to a combination of all these
conditions. Under such circumstances an effort
should be made to cleanse the stomach as rapidly
as possible by an even freer use of hot water, limit-
ing to a greater degree the quantity of food taken,
and insisting on more physical and mental rest.
At the same time the nerve tone should be improved
as rapidly as possible by tonics, massage, and electricity. But, inasmuch as patients differ and diseases vary in severity, it is easy to understand that
set rules in regard to the quantity and temperature

of water, or to the amount of animal food to be
given, cannot be laid down. The amount of hot
water should be sufficient to maintain the specific
gravity of the urine at about 1.014, and the quantity of meat should be as much as can be digested.
Whether the patient is drinking enough or is digesting his food properly is to be decided by the
physician and never by the caprice of the sick one.
It is to be borne in mind that the object of the use
of hot water and a strictly animal diet is to prevent
excessive fermentation, which is the underlying
cause of the diseased condition, and therefore it
should be employed systematically and persistently.

It is alleged by some, who are ignorant of facts,
that this single article of diet will bring on dyspepsia, Bright's disease, and other serious troubles, and
that it tends to establish a sort of meat habit, so
that the organs of digestion will not tolerate other
kinds of food. I will not take time to discuss asser-
tions and theories, but simply say that, in the treatment of hundreds of cases according to the methods
here given, I have never seen any evil results. On
the contrary, patients are gradually brought around
to a mixed diet as soon as safe for them ; the great
majority get well, so that they can eat a reasonable
amount of any kind of food, and in old chronic cases
of twenty or thirty years' standing they are made
comfortable and able to eat all that is necessary to
supply the requirements of their system.

Having thus far dwelt on the general principles
of treatment, I will now speak a little more specifically of the treatment of functional diseases of the
stomach. In cases of this nature the patient should

be held very closely to some form of animal food—
such as the muscle pulp of beef, beefsteak, lean
mutton, white meats of fish and poultry, the pulp
of oysters, well-fried bacon, and soft-boiled or
poached eggs. The prepared muscle pulp of beef
may be used, but more to furnish variety than because really essential. But, as a rule, these cases
will do better if the physician advises an almost
constant use of either broiled or roast beef or mutton, eaten slowly and thoroughly masticated before
being swallowed. It is also well to allow a very
small quantity of starchy food, in the proportion of
three or four parts of animal to one of starchy food
by bulk. It is safest to advise a very small piece of
dry toast—so dry that it will snap—or a piece of
stale roll, or a small piece of stale bread, or a tablespoonful of well-cooked rice or cracked wheat
dressed with butter, salt, and pepper. In the matter of vegetables it is well to advise a few tender
sprigs of celery, a little watercress, or a little horseradish, prepared with lemon juice instead of vinegar.
Moreover, the patient should be directed not to
swallow coarse particles of any of the substances
named, and to eat a moderate quantity and very
slowly. The drinks to be allowed at meal times
are a single after-dinner cup of black tea or black
coffee, sweetened with saccharin if desired. If
these are not well borne a cup of hot water, flavored
or not with lemon juice, may be taken. If the
functional cases are at all recent and these precautions are observed, it will require but a few days to
show a marked difference in the fermentation and
in the comfort of the patient. As soon as the un-

comfortable feelings have disappeared, the products
of fermentation eliminated from the blood, the
symptoms and physical signs of fermentation gone,
and as soon as the urine shows normal characteris-
tics, being absolutely free from biliary coloring matter, the patient may be given a larger proportion of
starchy food—say, one of starchy to two of animal
food. In functional cases of stomach and bowel
disease patients are to be kept on this routine as to
food and drink for a few weeks or months after
the evidences of excessive fermentation have ceased.
At the end of this time the patient may be led up,
little by little, to other food, such as fresh garden
peas, string beans, half of a baked potato, and a
few peaches, prunes, or grapes. These should be
given in small quantities at first, and if they cause
any trouble they should be discontinued and re-
course had to a rigid animal diet until the digestion
has returned to a normal state. And here let me
say that it is surprising how little gas is contained
in the intestines of people whose digestion is absolutely healthy. It is equally surprising to note
the serious disturbance of the mucous membrane
after a few weeks or months of excessive fermentation. On the one hand I have seen cases that have
given evidence that fermentation had existed in excess for twenty or thirty years without perceptibly
affecting the general health. On the contrary, I
have seen many cases where the most serious structural changes had resulted after only a few weeks
of indigestion and fermentation, either in the stomach or bowels, or else in both. It can only be said
in explanation of this that one is endowed with

great resisting power, while the other is not so blessed. In other words, these conditions work but little injury in robust persons, while in others of less resistance and stamina they may cause decided damage and great suffering. Therefore I do not put very great stress upon fermentation and gas when they occur in people of good health ; but they do have a very decided meaning when the health begins to fail and there are indications of serious structural change in the mucous surfaces of the stomach or bowels.

In the more strictly catarrhal states of the stomach or bowels, or of both, their lining becomes coated with an excess of sour, offensive, adherent mucus.. This material is in a large degree a ferment, and, as a consequence, sweet and starchy foods are soon transformed into a sour, yeasty, irritating, and injurious liquid. If this state of things is long continued it almost inevitably causes either vomiting of highly acid irritating liquids, or else frequent discharges from the bowels of gaseous products, undigested food, and thick, stringy, gelatinous mucus. The mucus thus cast off may be like the white of an egg, only more yellow ; or a thin, black, gelatinous substance ; or a thin, stringy material resembling wet tissue paper ; or, lastly, a distinct membranous exfoliation. It is in cases of this kind that an abundance of hot water, long continued, is of the highest utility for washing out the products of fermentation and keeping the surfaces in a fit condition for digestion and absorption. This practice must be continued and persevered in for months and years before the alimentary tract becomes thoroughly cleansed and restored to the power and function of normal digestion. In catarrhal cases of the stomach the best food is the muscle pulp of beef, prepared in accordance with the methods described, and given as the patient is able to digest. It is well to hold patients on this diet from one to three months, because it is the only one that can be anywhere near perfectly digested. Later on other foods can be resumed, but with great caution. Among the first foods to be given should be fresh garden peas, string beans, fresh warm milk from the cow, a little tomato, or a few prunes, peaches, or grapes. The foods allowed at first should be guardedly chosen and taken in a cautious manner. Little by little the diet should be extended until ordinary diet can be taken with comfort. If at any time the patient shows signs of not digesting his foods, he should be brought back at once to the rigid animal diet and held there until the organs again do perfect work. If the patient is thus promptly and strictly returned to a restricted animal diet, he will be all right in a few days. On the contrary, if he is not so treated the former manifestations of disease will occur. It is impossible for a person ever to get so well but that, if he becomes sick again, it will be the weak organs that are assailed. Like causes will certainly produce like effects. It is to be borne in mind that the mucous membrane, while it may be sound, is still delicate and sensitive, and must be restored and strengthened up to its natural state. And if you consider particularly the changes that have taken place in advanced cases, not only in the mucous membrane but in the con-

nective tissue, glands, and sympathetic nerves, it
stands to reason that a good condition must be kept
up long after the evidences of the disease have disappeared. And I might add in this connection that,
in my experience, it takes from one to three years
to bring about such changes and to cure a catarrh
of the stomach and bowels.

In catarrhal disease of the bowels much the same
line of treatment is to be followed. The use of hot
water and the rigid animal diet must be persevered
in until all traces of the disease have disappeared
from the blood, the urine, and the fæces. After
this system of alimentation has been persevered in
thus long, there may be a very gradual return to the
vegetable and starchy regimen already defined. At
times there may be slight relapses, but these will
be readily corrected by a return to a rigid use of the
hot water and animal diet for a few days. But,
despite drawbacks, there will be a prompt resto-
ration of comfort and a gradual progress toward
health until recovery is complete.

The extent to which I have gone into the general
principles of treatment and diet may lead to the be-
lief that I am indifferent to the place and power of
medicines in dealing with the functional and catarrhal diseases of the stomach and bowels. Such,
however, is not the case, for there are medicines of
very great utility and upon which I have come to
rely with confidence. In my judgment there are
four leading indications for the use of medicines in
these cases. There is a need for those that supplement the gastric juice, that stimulate the appetite,
that invigorate the nervous system, that excite or

retard the secretions, and that bear upon complications which may arise.

1. Among those of the first class is to be named pepsin, which is especially useful in aiding the digestion of animal food. To this useful agent may be added either bismuth, ginger, or ipecac, as needed.

2. Of the medicines calculated to stimulate the appetite I have found benefit to result from the preparations of cinchona, gentian, fluid extract of stillingia, and Fothergill's antidyspeptic pills.

3. In cases that need a decided nerve tonic to invigorate a feeble nervous system, and especially the nerves supplying the organs of digestion, there is nothing more advantageous than the preparations of strychnine and damiana.

4. For remedies to regulate the secretions I have obtained good results from the guarded use of Carlsbad salts, compound licorice powder, fluid extract of cascara, mild laxative pills, and hydrastin. On the contrary, when secretions become too free I often prescribe mild tonic astringents, like the fluid extract of blackberry root, fluid extract of hamamelis, bismuth, or chalk mixture. In catarrh of the stomach, duodenum, or bowels a combination of hydrastin and bismuth has rendered most excellent service. Hydrastin and bismuth seem to exert a peculiar and salutary effect upon mucous surfaces. In those cases where there is a marked tendency to acidity and fermentation, salicin alone, or with bicarbonate of sodium, or charcoal and magnesia, have given good and prompt results. Salicin usually affords excellent results, because it

does not disturb the stomach, is tonic in its action, and is one of the best agents we have to counteract acidity and the evils of fermentation. There are, of course, many other remedies to be used in the treatment of the diseases under consideration, but these are the principal ones which, if properly prescribed, are of great service.

In conclusion, there are two features in the clinical history of cases treated after the manner here outlined that are worthy of special note. In the first place, there is a natural tendency for the patient to gradually get weaker and thinner. The deprivation of starch, sugar, and fat cuts off, so to speak, the "kindling wood" of the system that affords immediate strength and heat. Not only that, but excessive fermentation, especially alcoholic, is to some extent a stimulant, and it is the loss of its chronic effects that is felt by the system. No injurious consequences follow this weakness, however, if the patient believes what has been told him and does as advised. After a time the blood becomes richer, the nervous system stronger, and renewed strength takes the place of former debility. If necessary, as a matter of bridging over temporary weakness, the patient can be given from a teaspoonful to a tablespoonful of old whiskey or brandy, in water, from one to two hours after meals.

And, secondly, in catarrhal cases of the bowels where the movements occur several times daily, it is sometimes necessary to bring them under control with simple remedies, like external heat, rest in bed, and the internal use of mild doses of bis-

9

muth, chalk mixture, or fluid extract of ginger.
These rather frequent movements, while they sometimes weaken, are, after all, salutary. They are,
as it were, Nature's "house cleaning," removing
the products of fermentation, exfoliations of mucus,
and other morbid material.   Before or during these
frequent movements or clearings the patient may
experience local or general muscular or neuralgic
pains, but all of these temporary disturbances soon
pass away.

Such, then, are the methods which, in my judgment, are the best of all for removing the causes
of functional and catarrhal diseases of the stomach
and bowels, restoring the quantity and quality of
the blood, augmenting the force of the nervous
system, and putting the general health on a solid
basis.

As you have already heard, they consist—

1. In securing a willing, obedient, hopeful, and
confident mental condition.

2. In making a careful diagnosis, based on the
usual methods, and, in addition, a frequent microscopical and chemical study of the products of the
system, as the blood, the urine, and the fæcal discharges.

3. In placing the patient under the most favoring
hygienic conditions.

4. In an intelligent and systematic use of hot
water for the purpose of cleansing the surfaces of
the stomach and bowels, stimulating the secretory
and excretory functions of the liver, kidneys, and
other glands, and supplying the system with the
requisite amount of liquid.

5. In using an article of diet that undergoes but slight if any fermentation, that can be easily digested, absorbed, and assimilated, and that will make, in time, the maximum amount of blood and nerve force. The great object is not to arbitrarily put the patient on a particular article of diet, but rather on one that will meet the above-named requirements and tide him over until well enough to resume the use of various articles. For this purpose I have not found any food comparable to the muscle pulp of beef, prepared and used as before described. To afford the greatest service it must be carefully prepared, properly eaten, and thoroughly digested. To know whether it is well digested, reliance must be placed on the usual signs and symptoms, and on a frequent microscopical and chemical study of the blood, the urine, and the fæces. The latter method affords the most accurate means of determining what manner of work is being done in the laboratory of the system.

6. In the use of medicines in so far as they improve the appetite, excite or retard secretions, restore the blood and nervous system, and meet varying conditions and complications, if any develop.

II.

On the Nature and Preventive Treatment of Seasickness.

Nowadays inventive genius and the progress of science have made travel by sea rapid and safe. The great steamers pass quickly and triumphantly against wind and wave from point to point and from shore to shore. The world is made smaller, nation is drawn closer to nation. Seasickness is the chief barrier that remains; it is the almost certain affliction of those who use this mode of travel, be it for health, pleasure, education, or the purposes of trade. This peculiar form of vertigo it is that Neptune imposes as a tax on all of his subjects, except a favored few. It is estimated that only about three per cent of all sea-goers are exempt.

Mechanical science has very materially shortened the duration of the disease by increasing the rapidity and comforts of travel. The layman has pretty thoroughly discussed the subject, and seems never to tire when considering its humorous side. The medical profession has done very little, and written and thought less. It is with the desire to excite serious study of this neglected disease that this article is written. No effort is made to discover "some new thing"; no claim will be made for originality. The united thought of the profession may be able to lift the cloud that obscures the nature of

the trouble, and devise some means for its prevention or alleviation.

On account of the nature and limited adaptability
of our organism, which is fitted, by creation and
habit, to life on the stable and solid land; on account
of the great change in the environment when on
the restless sea, it is folly to hope that the evil can
be wholly overcome. So long as the rolling and
pitching ship is at the mercy of every wave, and,
impressing its restlessness on every object that can
be felt and seen, takes from us the guides and gov-
ernors of co-ordination and of equilibration : so long
as these disordering and uncorrected sensory impressions possess correlatives in consciousness, the
vertigo of mariners will be produced. For seasickness is essentially and primarily a disordered sense
of equilibrium and of space, a sensory form of vertigo.

The symptoms and their order and manner of development confirm this view. The first and essential sign of every case of seasickness is a feeling of
dizziness or lightness of the head, or vertigo. It is
the most invariable, and the most persistent, and
sometimes the only symptom. It is alone present
in the prelude; though overshadowed, is never absent from the scene ; and is the last to leave the
stage when the curtain falls. It is commonly associated with headache, an indefinable nervousness,
sensitiveness to light, a contracted pupil, and a
keen sense of smell. The temper is extremely irritable, the face is flushed or pale, or rapidly changes
from the one to the other state—the vaso-motors
and inhibition are struggling for the mastery. The

condition is one of hyperæmia and instability of the
sensory and sympathetic nerve centres. These epi-
phenomena may be absent and the voyage completed with only varying degrees of vertigo. But
more often the simple vertigo is followed by nervous exhaustion and mental depression, muscular
inco-ordination and relaxation, a weak heart, low
arterial tension, salivation, nausea, and vomiting.
The irritability of weakness supplants the sensory
excitement, and the vertigo is increased by the cerebral anæmia.

Thus we have three pretty well defined forms or
degrees of seasickness—sensory vertigo, sensory
vertigo with cerebro-spinal irritability, and vertigo
with prostration.

The form and degree and duration of the attack
depend on the nature and intensity of the movements of the ship, on the susceptibility and adaptability of the individual, and the incidence of the
disturbance. When the cerebro-spinal system is
most involved, vertigo, headache, and nervousness
are marked; when the sympathetic is weakest, the
nausea, vomiting, and prostration are most prominent.

The nervous irritability may be explained as the
result of the cerebral excitement and the uncommon
and numerous sensory impressions. The cerebro-
spinal hyperæmia is due partly to the increase of
functional activity, and partly to the tonic contraction of all the muscles driving the blood out of the
musculo-venous reservoir. Every peripheral excitation determines neural discharges and causes an
augmentation of potential energy. It is also well

known that the pupil contracts under the influence
of exciting sensations, as does also the whole reflex
muscular system.

The vomiting, in the popular mind, constitutes
the essential part of the malady. Many physicians,
it must be admitted, adopt this idea and embody it
in their treatment. Now, we would state with emphasis that acute dyspeptic attacks must not be
confounded with seasickness. Acute dyspepsia is a
powerful predisposing cause of the disease, but has
no relation whatever to the movements of the ship.
The cause must be sought in overeating, irregular
habits, loss of sleep, overwork, worry, anxiety,
grief, the abuse of drugs—in some gross violation
of the hygiene of digestion. The disturbance of the
stomach is primary and would have occurred under
similar circumstances on land. The vomiting of
seasickness seems to be the effect of the cerebral
anaemia produced by the weak heart, vaso-motor
disturbance, and muscular relaxation—all due to
paresis of the sympathetic from fatigue of the nerve
centres by sensory overexcitation, or from emotive
shock, or from excessive inhibition through a sense
of defective motor innervation and of failure to
preserve the equilibrium of the body.

From this analysis it will appear that the symptoms referable to the nervous system are primary
and controlling, and that the essential sign of seasickness is vertigo. This, then, limits the explanation to the production of the vertigo by the ever-
varying and complicated movements of the ship,
for all observers agree that this is the remote cause.
How is the vertigo produced ?

The process is not a simple one. Many theories
fall short of the mark because they do not include
enough ; because it is incorrectly assumed that only
one line connects the cause with the effect. It is
my purpose to show that the motion of the ship is
connected with the vertigo by many routes that
the mechanical cause splits up and reunites in the
biological effect. On the one hand we have the
movements of the ship, and on the other are the
disturbed sense of equilibrium and of space manifest in consciousness as vertigo. How, then, do the
movements of the ship disturb these two senses in
this peculiar manner ?

It is foreign to our purpose to discuss the nature
of the sense of equilibrium, whether it be the corre-
late in consciousness of afferent sensory impressions
or a central sense of motor innervation. Nor would
anything be gained by disproving the existence of
so-called spinal and muscular perception. It is the
reality and composition, and not the location, of the
sense of equilibrium with which we are concerned.
The sense of equilibrium is a compound one and is
correlated in consciousness with many peripheral
impressions—muscular, tactile, labyrinthine, visual,
and from pressure. Through the muscular we are
cognizant of the state and position of a part as related to the rest of the body. By the other sensory
impressions we are informed as to the relation of
the body to surrounding objects and to the vertical
position. Now, the perfection of the sense of equilibrium is dependent on the integrity of the sensory
impressions which compose it. When the information is false or falsely interpreted the motor inner-

vation will be wrong and the result bewildering.
When the perception of relations is incomplete and
deceptive and uncorrected, there result inco-ordination and unsteadiness and vertigo.

The disordered sense of equilibrium is sufficient
alone to produce the vertigo of mariners, for the
blind are not exempt. Deafness seems to confer a
certain degree of immunity, and closing the eyes
will often diminish the vertigo. It is through the
sense of sight and the perception of the muscular
changes of convergence and divergence and accom-
modation that the sense of space is built up. Insufficiency and inco-ordination of the ocular muscles often give rise to vertigo. It is through the
eye also that we are chiefly made cognizant of our
position in space. Where the perceiving subject is
in motion the false perception of relations is projected outward as an illusion of moving objects.
The subjective feeling of this disorder is vertigo.
The dizziness of high altitudes and openness or void
arise from a disordered sense of space.

Vertigo may be divided into three large classes :
It may be cardio-vascular, as the vertigo of cerebral
anæmia or of arterial sclerosis ; it may be of central
origin, as the vertigo of properly located brain tu-
mors ; or it may be the peripheral or sensory form,
of which the vertigo of Ménière's disease and seasickness may be taken as a type. We have already
stated that the vertigo of seasickness with prostration is partly due to cerebral anæmia, or, in
other words, is also cardio-vascular. But the essential and primary vertigo is of a purely sensory
origin.

The preservation of equilibrium is dependent on :
(1) the integrity of afferent impressions ; (2) on
proper motor innervation guided by past experience,
and grouped and limited so as to produce a purposive movement or maintain a definite relative
position : (3) on proper muscular response, (4) the
result of which is reflected to the co-ordinating and
higher centres, and there is appreciated as efficient
or defective. When on an irregularly moving body
none of these conditions can be realized, and on
board a ship, in a rough sea, the difficulty may be
insurmountable. The sensori-motor nerve circuit
carries within itself the power of co-ordination without the connection or intervention of the higher
centres, though the higher centres may regulate or
control. Equilibration is commonly an unconscious
process. We are not conscious of all the peripheral
impressions which are co-ordinated into vertiginous
movements; we merely have a sense of the defective motor innervation. The defect, the discord,
the false association, the confusion of relations, are
felt as vertigo if they rise into consciousness or are
not displaced by a more potent feeling.

With these explanations turn we now to the consideration of the manner in which the senses of
equilibrium and of space are disturbed by the movements of the ship as it pitches or rolls or mixes the
two motions. The body is constantly thrown out
of equilibrium, and the position of the surface which
supports it cannot be appreciated. The sensations
of contact and of pressure ever vary in degree and
in direction—now slight as the ship sinks, the individual feeling as if left in midair : now great as

the ship rises and presses against the descending
body. The same uncommon and confusing sensory
impressions arise also from the movable viscera and
internal sensory surfaces, particularly from the
semicircular canals, through oscillations of the endolymph or hyperæmia of the auditory centre—sensations associated in experience with other positions
of the body than that which it now occupies. No
change in movement can be anticipated ; no position of the body can be thoroughly made out. The
sense of sight cannot be utilized to correct and
guide—an ever-changing point of view amid ever-
changing objects ; all the sensory impressions which
make up the life of relations are bewildering. The
fault does not lie in perception, nor in co-ordination,
nor in the periphery—the sensory mechanism works
perfectly. It is because the sense of want of support and the other peripherally excited afferent impressions are disordering. It is because new sensations, from an environment to which the organism
is not adapted, obtain a false association in consciousness. It is because relations cannot be made
out as they really are ; because the erroneous inferences as to the relations of the body to objects seen
and felt are out of harmony with the other sensory
impressions ; because the results of the efforts to
maintain equilibrium cannot be verified. And the
central confusion and discord and false association
are projected into the outer world as illusions of
movement and of space—a simple disorder of relations—a sensory form of vertigo. Such seems to be
the explanation of the vertigo which is the cardinal
sign or synonym of seasickness.

There are few subjects at once so unsettled and so speculated about as the causation of seasickness. It is not contended that the view here set forth is complete and final. But it is believed to contain the germ of the truth, and is based on the study of the symptoms in the light of physiology and pathology. It best explains all the phenomena, and the cause acting in the manner indicated will produce the vertigo to which, and to the condition of the cerebro-spinal and ganglionic nerve centres, all the symptoms are sequential.

It may be of interest to mention briefly and in the order of their publication the theories which at different times have commanded the most consideration and credence :

1. It is due to fear (Plutarch), proof of which is that infants who cannot reason, and animals, are exempt (Gerépratte).

This theory is only interesting because it still survives in the pretty widespread relief that the development of seasickness can be influenced or prevented by the exercise of the will and a mental attitude of indifference. Nothing can be more ludicrous than a traveller trying to ward off seasickness by force of will, unless it be a philosopher striving to suppress a toothache, or a poet to charm away the gout by the power and sweetness of his song. Strong feelings and powerful emotions can temporarily supplant in consciousness the sensation of vertigo. Animals are not exempt, though they do not vomit. The cause alleged is inadequate, and the evidence is made up of false observation.

2. It owes its existence to sympathy between the

brain and peripheral nerves disturbed by the movements of the ship (1756, Gillchrist).

In the early dawn of physiology this is a very shrewd guess.

3. It is due to cerebral congestion and irritation arising from minute concussions of the brain by the fluids of the body during the descent of the ship, analogous to the rise of the mercury as the barometer is dropped (1810, Wollaston)

Minute concussions would produce headache analogous to that from riding a rough horse, but not vertigo. The onset should always be gradual and slow. Slight movements should have no effect. A simple change in the character or cessation of the movements should never remit or inaugurate the trouble. The cause is inadequate, cannot be shown to be operative, and the blood vessels are fortunately not dead, rigid tubes. Infancy with its soft blood vessels, and old age with its hard arteries, are alike almost exempt.

4. It is produced by the influence of the visceral movements on the diaphragm (1824, Jobard and Kérandren).

Again the influence is inadequate. The symptoms are not reproduced or explained in the order of their development. And fixation of the viscera by an abdominal band exerts only a slight influence by diminishing the peripherally excited impressions.

5. The movements of the ship in an arc-like zigzag line arouse a centrifugal force which so influences the circulation in the aorta as to diminish the amount of blood going to the brain. The anæmia

of the brain results in cerebral depression, which through the sympathetic invokes vomiting. This author considers the vomiting a conservative process induced to supplement the deficient quantity of blood sent to the head (1847, Pellarin).

This is an exquisite use of "occult influences" and the reputed "beneficent purposes" of Nature.

6. It is intoxication by a marine miasm developed in the decaying animal and vegetable matter of the sea, and aroused from its hiding place during the agitation of the water by the ship or wind or wave (1850, Sémanas).

If this theory were fresh from a bacteriological laboratory it might command nowadays a great deal of consideration. It was based on a false analogy. But the large doses of quinine recommended may be of benefit by producing anæmia of the semicircular canals (if this condition be true).

7. The proximate cause of seasickness is the heaping of the brain mass upon itself by centrifugal force, and subjecting the part to pressure against the bony casement, or to the hurtful centrifugal movements of the cerebro-spinal fluid, which also leave parts of the brain exposed to injury. Preference is given to the latter view (1856, Fonssagrives).

This is a further stage in the development of the mechanical theory, which is fast approaching an absurdity.

8. The proximate cause is hyperæmia of the spinal cord, especially in those segments related to the stomach and muscles concerned in vomiting, induced directly or reflexly by the irritating movements of the brain, spinal cord, abdominal and pelvic viscera,

and by jerks on the spinal ligaments. The involun-
tary muscles are disturbed by the unwonted number
of impulses transmitted to them from the preter-
naturally excited spinal cord (1864, Chapman).

This theory marks the beginning of a new era.
A good many threads of truth run like gold through
the dark web, and physiology is in an able manner
brought to the aid of the old theories of small concussions and mechanical irritations. The treatment
by means of the spinal ice-bag does not seem to have
increased the comfort of travellers.

9. It seems to be due to the sudden and recurring
changes of the relations of the fluids to the solids of
the body (1868, Barker).

10. It is due to the disordering movements of the
cerebro-spinal fluid, from which results an intermittent anæmia and a certain degree of commotion
of the cerebral mass. Children are exempt through
expansibility of the fontanelles (1868, Autric).

It does not seem plausible that a force sufficient
to cause the fontanelles to bulge would not compress
the very yielding blood vessels of childhood, and
children with widely open fontanelles are not always
exempt.

11. It is due to the continued action on the brain
of a certain set of sensations, more particularly the
sensation of want of support (Carpenter, Bain, and
(1872) Pollard).

This is a development of the very shrewd guess of
Gillchrist. It stands at the beginning of new views.
The mechanical theories do not seem to have gone
much beyond "possibilities" in their explanation
of the symptoms. Experiments, observed order

of sequences, and logic now turn on a flood of light.

12. Seasickness is a functional disease of the central nervous system, mainly of the brain, but in some instances of the spinal cord also, the result of a series of mild concussions (1880, Beard).

The cause is inadequate, and functional disease of the central nervous system is not very definite or lucid. The preventive treatment by bromization, however, was a great advance in therapeutics.

13. Motion produces sickness by disturbing the endolymph in the semicircular canals, the viscera in the abdomen, and possibly the brain and subarachnoid fluid at its base (1881, Irwin).

14. All the symptoms of seasickness can be explained by paresis of the sympathetic (1887, Skinner).

This is a very important factor, but how is the paresis induced? It is an epiphenomenon, and an important indication in the drug treatment.

15. Vertigo and vomiting are the essential symptoms. The movements of a ship in a storm, particularly its quick descent, cause movements of the cerebro-spinal fluid, and cerebral blood is displaced and the brain subjected to shocks and the cerebellum to commotion ; or movements of the abdominal viscera and contractions of the diaphragm, with their resulting local action and reflex inhibitory influences (1888, Pampoukis).

16. The symptoms of seasickness are those of cerebral anaemia. The uncommon and disordering movements that are felt derange and diminish reflex muscular tonicity and contraction, which maintain

equilibrium and regulate the return venous circulation. Then results a muscular relaxation, of which
the loss of equilibrium is the sign and the cerebral
anæmia the consequence (1890, Rochet).

It seems that too great prominence is given to loss
or diminution of reflex muscular tonicity. Fatigue
is chiefly central, and the most highly endowed and
the most differentiated tissue is the first to become
exhausted. We have seen that in the production
of the paresis of the sympathetic and prostration
central fatigue is one of the factors. It seems that
muscular relaxation would have to be pretty well
marked before there could be much interference
with the return venous circulation. And vertigo is
present when the pupil is contracted under exciting
sensations and the traveller is walking in the dark.
The theory makes a deferred result the active cause,
but withal is the best explanation yet given.

There are varying degrees of susceptibility to the
disease. We have seen how powerful a predisposing cause is acute dyspepsia. The anæmic, the
neurotic, the neurasthenic yield very readily to it,
as do all who have weak and easily excited nerve
centres. Athletes in training have been prostrated,
while delicate women were laughing at their discomfort. Infancy and old age are more exempt
than middle life. Individuals subject to vaso-motor
disturbances are predisposed to the malady. All the
symptoms have been often reproduced on land, after
the lapse of months, by association of ideas.

Seasickness is not a fatal disease. Deaths have
been recorded as due to it, but in these cases it only
caused the already suspended sword to fall. Sea-

10

sickness is an evil; it is never " very good at times "
(Burton), nor " salutary " (Johnson). All the good
effects of sea travel are obtained without it. It is
a dangerous malady when organic disease of the
heart or blood vessels, or of the stomach, or of the
nervous system, or of the lungs, liable to be attended
by hæmoptysis, is present. It nearly always delays
or disorders menstruation, and, as is well known,
has often terminated pregnancy. It sometimes
persists for a variable period after the voyage, and
some never completely recover their sense of equilibrium and of space.

Bad treatment is the natural sequence of false
views of causation. When we know how a symptom or disease is produced the management becomes
rational, though not always efficient. To the consideration of the preventive treatment a few practical
suggestions will be added on the management of the
attack.

In the prevention of seasickness we work along
two lines—the removal of the predisposing causes and
the diminution of the action of the exciting ones.
In each instance we strike at causation, and the
effect of the double blow is commonly satisfactory.
My attention was first drawn to this method of
prevention by the comparative immunity from seasickness of patients who were under my treatment.
before and during the voyage, for some one of the
many disorders and diseases of nutrition. So far
my experience with the method has not been very
great, only a few more than one hundred cases having been managed in this manner. The number of
cases is only large enough to suggest rather than

establish the value of the treatment. But if it be
understood that more than half of these travellers
had been previously so sick that they turned with
horror from the repetition of the voyage, and that
more than three-fourths of them completed the passage under the influence of my method without the
slightest qualm, and subsequently, when neglect-
ing my directions, became fearfully ill, it may be
thought advisable to state the method to the profession with a view to having its utility thoroughly
tested.

The treatment as directed to the digestive system
has one important object in view—to diminish the
irritability of the sensory-nerve endings of the mucous lining of the alimentary canal by keeping the
digestive tube functionally active, clean, and sweet,
and the consequent prevention of acute dyspeptic
attacks. And we follow up the advantage thus
gained by securing active elimination and perfect
assimilation and disassimilation, thus strengthening
and saving from the irritation of an impure blood
the nerve centres, whose overexcitation and fatigue
play so important a rôle in the development of the
malady. In a few words, we strive to promote a
high degree of healthy nutrition, because we believe
that a strong man is best prepared to resist the
encroachments of disease. Good nutrition is a well-
fitting armor that turns aside many a deadly blow.
If we succeed in realizing this high endeavor, I do
not believe that the anæmic stage of seasickness
will be developed.

Close attention to the hygiene of nutrition will
enable us to get the vital processes on a physiologi-

cal basis. Only a few days will be required for this
purpose if there be but slight disorder of one or
more of the nutritive processes. The week before
sailing is commonly one of excitement, dissipation,
and worry. All preparations for the voyage should
be completed several days before going aboard—
the bowels regulated by laxatives, the secretions
righted and supplemented if requisite, elimination
keep free, and a plain, easily digested, and easily
assimilated diet should be adopted. In a general
way the sweets and starches should be limited, and
lean meats made the staple food. But the age, activity, peculiarities, habits, the needs of general nutrition, the capability of the digestive organs, must
all be taken into consideration in the selection of the
diet. The means must be varied to suit each special
case, for individualization is the secret of success.
But the aim is simple and definite—to secure the
perfect digestion and assimilation of a sufficient
quantity of food to meet the requirements of nutrition. If the patient gets eight hours of restful sleep
every night, and feels no pain or discomfort or
drowsiness after meals ; if there is no flatulence ; if
the urine contains no abnormal coloring matter nor
excess of phosphates, urates, or uric acid, and the
stools are normal—we know that the food is being
digested, absorbed, and assimilated in sufficient
quantity, if there be no loss of strength to meet the
demands of life, and that the excretory products
are changed into their simplest and most soluble
and most unirritating forms. Until this state of
nutrition is established the patient is not prepared
for the voyage. The same simple and regular and

temperate way of living and eating must be observed throughout the passage.

When there is a serious derangement or disease of the digestive system, the proper treatment must be instituted to secure the one aim of healthy nutrition. How this can be undertaken with the greatest hope of success has been outlined by me in articles published in the New York *Medical Journal*.

The second part of the preventive treatment is intended to diminish the activity of the exciting causes until the organism can adapt itself to the new environment and become inured to the disordering sensations.

During the first forty-eight hours it is advisable to remain in bed and sleep as much as possible. The effort to maintain equilibrium is diminished, the confusion through the sight of moving objects is limited, the life of relations is " cabined and confined," consciousness is diminished at last. Four light meals should be taken a day and very little fluid drunk. The danger of a mechanical hyperæmia of the nerve centres, by excessive muscular tonicity forcing the blood out of the musculo-venous reservoir, will be obviated. The only drink should be a single cup of hot water with each meal.

After the expiration of this preliminary period, during which the action of the exciting cause is weakened and the organism is being accustomed to the disordering sensations, the time, except that which is regularly given to sleep, should be spent in the open air on deck. The sensory vertigo which is ever ready to arise into consciousness must be sup-

planted by purposive movements, the efficiency of
which can be verified, as walking, etc., and by mental occupation or diversion.   It is well known that
intense fear or excitement or absorbing thought
will dissipate " the swooning sickness on the dismal
sea." The vertiginous sensation is driven out of
consciousness by the commanding presence of a
powerful emotion, feeling, or thought.

A widely known method of diminishing the action of the exciting cause is by the use of the bromide of sodium, which must be pushed to its full
physiological effects and the influence kept up during the entire voyage.   The neuro-muscular disorder is controlled, and sensory perception, both peripheral and central, is dulled.   The drug influences
favorably the simple vertigo, prevents the development of the hyperæmia, but it intensifies the misery
of the anæmic form.   The treatment is often efficient, but it should never be tried except on the
advice and under the supervision of a physician.
Seasickness itself is not so harmful as may be bromization.   The large doses usually upset the stomach, and the drug irritates all the organs by which
it is eliminated.   The bromides, when pushed to the
point of poisoning, often exert a persistent and pernicious influence on the nervous system.

The treatment during the attack is quite different
in the anæmic and the hyperæmic varieties.

When hyperæmia is present the influence of the
exaggerated reflex muscular tonicity can be diminished by voluntary muscular movements, which require muscular relaxation as well as contraction for
their performance.   The vertical position is an advantage.   A hot foot bath will also draw the blood

away from the nerve centres, as keeping the feet in very hot water for some time has produced syncope. A very powerful effect can be produced by placing the hands and feet in hot water and applying ice to the head and spine. Counter-irritation is a procedure of questionable utility. Caffein will suppress the sense of central fatigue. Antipyrin or bromide of sodium by the rectum may be of some use.

In the anæmic stage such drugs as must be absorbed to produce an effect should be given hypodermically. Atropin is the best drug to stimulate the paretic sympathetic, but nitroglycerin must be given simultaneously to dilate the arterioles. Strychnin and the natro-benzoate of caffein also meet obvious indications. Ergotin, on account chiefly of its action on the urine, is also valuable. The judicious use and combination of these five remedies will meet the indications from the side of the muscular, nervous, and circulatory systems. Whiskey (and food also) may be required by the rectum. The horizontal position, with the head low, should be persistently maintained. The vomiting will also be favorably influenced by the preceding drugs. Copious draughts of hot water, to wash out and soothe the stomach, is a remedy of great value. Frequently repeated and small doses of creosote, with lime water and an infinitesimal quantity of ipecac, may be effectual. Oxalate of cerium, in five-grain doses every hour for three or four administrations, is another good remedy. If these preventive precautions and remedies fail, the patient must content himself until he can again get into his element, the place where he was created and educated to live—on land.

www.ingramcontent.com/pod-product-compliance
Lightning Source LLC
Chambersburg PA
CBHW021811190326
41518CB00007B/552